AF360818

New Research Advances on Marine Invertebrates

New Research Advances on Marine Invertebrates

Editor

Alexandre Lobo-da-Cunha

MDPI • Basel • Beijing • Wuhan • Barcelona • Belgrade • Manchester • Tokyo • Cluj • Tianjin

Editor
Alexandre Lobo-da-Cunha
University of Porto
Porto
Portugal

Editorial Office
MDPI
St. Alban-Anlage 66
4052 Basel, Switzerland

This is a reprint of articles from the Special Issue published online in the open access journal *Journal of Marine Science and Engineering* (ISSN 2077-1312) (available at: https://www.mdpi.com/journal/jmse/special_issues/F_Marine_Invertebrates).

For citation purposes, cite each article independently as indicated on the article page online and as indicated below:

LastName, A.A.; LastName, B.B.; LastName, C.C. Article Title. *Journal Name* **Year**, *Volume Number*, Page Range.

ISBN 978-3-0365-6658-0 (Hbk)
ISBN 978-3-0365-6659-7 (PDF)

Cover image courtesy of Rita Azeredo

Contents

About the Editor

Alexandre Lobo-da-Cunha

Alexandre Lobo-da-Cunha is a full professor of the University of Porto (Portugal) who is mainly interested in the study of marine molluscs, using, among other research methods, light and electron microscopy to investigate the digestive system, gills and other organs of these animals. He also participates in microbiology studies by applying transmission electron microscopy.

Journal of
Marine Science and Engineering

MDPI

Editorial

New Research Advances on Marine Invertebrates

Alexandre Lobo-da-Cunha [1,2]

1   Department of Microscopy, Institute of Biomedical Sciences Abel Salazar (ICBAS), University of Porto, Rua de Jorge Viterbo Ferreira, 228, 4050-313 Porto, Portugal; alcunha@icbas.up.pt
2   Interdisciplinary Centre of Marine and Environmental Research (CIIMAR), Avenida General Norton de Matos, Terminal de Cruzeiros do Porto de Leixões, 4450-208 Matosinhos, Portugal

Citation: Lobo-da-Cunha, A. New Research Advances on Marine Invertebrates. *J. Mar. Sci. Eng.* **2023**, *11*, 6. https://doi.org/10.3390/jmse11010006

Received: 6 December 2022
Accepted: 8 December 2022
Published: 20 December 2022

Marine ecosystems encompass a wide variety of invertebrates, pelagic and benthonic, from intertidal to deep-sea habitats in polar to tropical regions. The invertebrates constitute a paraphyletic group of animals that includes all metazoans except those belonging to the chordate subphylum Vertebrata. Therefore, they are distributed through all the phyla of metazoans (a total of 36 phyla according to some recent classifications), including the Cephalochordata (lancelets) and Urochordata (tunicates) of the phylum Chordata [1]. It was estimated that invertebrates correspond to more than 92% of all living marine animal species [2], ranging in size from microscopic organisms [3,4] to the giant squids. Invertebrates are major components of oceanic food webs, being involved in nutrient cycling. Suspension-feeders, such as bivalves, remove particles from the water column, thus improving local water quality, and reef-builders, such as corals, are ecosystem engineers that create habitats for many other marine species [2]. However, a significant number of these species are currently endangered by costal habitat destruction, overfishing, ocean acidification and other forms of pollution that threaten marine biodiversity [2]. In addition to their high environmental importance, some marine invertebrates also have great economic value. Several species of crustaceans and molluscs are greatly sought after for human consumption, being produced in large quantities in aquaculture facilities or captured through fishing activities [5,6]. Moreover, the high diversity of marine invertebrates makes them a vast source of biomaterials and bioactive natural products with pharmaceutical, nutraceutical and antifouling applications, among other uses [7]. Nevertheless, despite all the research undertaken so far, much still remains to be investigated about these animals. Therefore, this Special Issue aimed to collect articles providing new and relevant information in this field. The seven articles included here reflect the diversity of marine invertebrates. The studies on echinoderms concern the histology of bizarre deep-sea holothurians [8], gut regeneration in holothurians [9] and the anti-cancer properties of the coelomic fluid of a sea-urchin [10]. Another article presents a morphological description of three different types of acellular material lining hemal spaces of a penaeid shrimp [11]. The study on Placozoa was focused on the crystal cells that have been implicated in gravity reception in these very small and simple animals that lack muscles and nerves, which feed by external digestion [12]. The article on the oesophagus, stomach and intestine of chitons disclosed new data about the histology and ultrastructure of the digestive system of these less-studied molluscs [13]. Finally, an extensive study on eunicid *polychaetes* revealed the influence that prostomial sensory organs have on brain microanatomy and showed that the specificities of the sensory organs support the latest phylogenetic relationships within this group of annelids [14]. In these articles, a variety of research methods were applied in order to progress our understanding of marine species, including light and electron microscopy, liquid chromatography, mass spectrometry, flow cytometry and cell and tissue cultures. With this Special Issue we also seek to stimulate further developments in the investigation on marine invertebrates, a field with great potential for fascinating new discoveries.

Funding: This research received no external funding.

Conflicts of Interest: The author declares no conflict of interest.

References

1. Edgecombe, G.D.; Giribet, G.; Dunn, C.W.; Hejnol, A.; Kristensen, R.M.; Neves, R.C.; Rouse, G.W.; Worsaae, K.; Sørensen, M.V. Higher-level metazoan relationships: Recent progress and remaining questions. *Org. Divers. Evol.* **2011**, *11*, 151–172. [CrossRef]
2. Chen, E.Y.-S. Often overlooked: Understanding and meeting the current challenges of marine invertebrate conservation. *Front. Mar. Sci.* **2021**, *8*, 690704. [CrossRef]
3. Carugati, L.; Corinaldesi, C.; Dell'Anno, A.; Danovaro, R. Metagenetic tools for the census of marine meiofaunal biodiversity: An overview. *Mar. Genom.* **2015**, *24*, 11–20. [CrossRef] [PubMed]
4. Kristensen, R.M. An introduction to loricifera, cycliophora, and micrognathozoa. Integr. *Comp. Biol.* **2002**, *42*, 641–651. [CrossRef] [PubMed]
5. Bondad-Reantaso, M.G.; Subasinghe, R.P.; Josupeit, H.; Cai, J.; Zhou, X. The role of crustacean fisheries and aquaculture in global food security: Past, present and future. *J. Invertebr. Pathol.* **2012**, *110*, 158–165. [CrossRef] [PubMed]
6. Mau, A.; Jha, R. Aquaculture of two commercially important molluscs (abalone and limpet): Existing knowledge and future prospects. *Rev. Aquacult.* **2018**, *10*, 611–625. [CrossRef]
7. Romano, G.; Almeida, M.; Varela Coelho, A.; Cutignano, A.; Gonçalves, L.G.; Hansen, E.; Khnykin, D.; Mass, T.; Ramšak, A.; Rocha, M.S.; et al. Biomaterials and bioactive natural products from marine invertebrates: From basic research to innovative applications. *Mar. Drugs* **2022**, *20*, 219. [CrossRef] [PubMed]
8. LaDouceur, E.E.B.; Kuhnz, L.A.; Biggs, C.; Bitondo, A.; Olhasso, M.; Scott, K.L.; Murray, M. Histologic examination of a sea pig (*Scotoplanes* sp.) using bright field light microscopy. *J. Mar. Sci. Eng.* **2021**, *9*, 848. [CrossRef]
9. Bello, S.A.; García-Arrarás, J.E. Intestine explants in organ culture: A tool to broaden the regenerative studies in echinoderms. *J. Mar. Sci. Eng.* **2022**, *10*, 244. [CrossRef] [PubMed]
10. Luparello, C.; Branni, R.; Abruscato, G.; Lazzara, V.; Sugár, S.; Arizza, V.; Mauro, M.; Di Stefano, V.; Vazzana, M. Biological and proteomic characterization of the anti-cancer potency of aqueous extracts from cell-free coelomic fluid of *Arbacia lixula* sea urchin in an in vitro model of human hepatocellular carcinoma. *J. Mar. Sci. Eng.* **2022**, *10*, 1292. [CrossRef]
11. Sidebottom, R.B.; Bang, S.; Martin, G. Microscopic anatomy of the lining of hemal spaces in the penaeid shrimp, *Sicyonia ingentis*. *J. Mar. Sci. Eng.* **2021**, *9*, 862. [CrossRef]
12. Mayorova, T.D. Differentiation of crystal cells, gravity-sensing cells in the Placozoan *Trichoplax adhaerens*. *J. Mar. Sci. Eng.* **2021**, *9*, 1229. [CrossRef]
13. Lobo-da-Cunha, A.; Alves, Â.; Oliveira, E.; Calado, G. Functional histology and ultrastructure of the digestive tract in two species of chitons (Mollusca, Polyplacophora). *J. Mar. Sci. Eng.* **2022**, *10*, 160. [CrossRef]
14. Kuhl, S.; Bartolomaeus, T.; Beckers, P. How do prostomial sensory organs affect brain anatomy? Phylogenetic implications in Eunicida (Annelida). *J. Mar. Sci. Eng.* **2022**, *10*, 1707. [CrossRef]

Journal of
Marine Science and Engineering

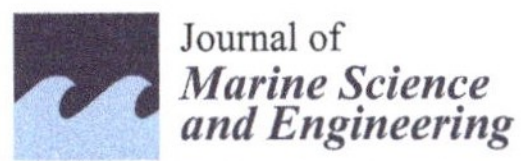

Article

Histologic Examination of a Sea Pig (*Scotoplanes* sp.) Using Bright Field Light Microscopy

Elise E. B. LaDouceur [1,*], Linda A. Kuhnz [2], Christina Biggs [3], Alicia Bitondo [3], Megan Olhasso [3], Katherine L. Scott [1] and Michael Murray [3]

[1]   Joint Pathology Center, 606 Stephen Sitter Ave, Silver Spring, MD 20910, USA; Katherine.l.scott12.mil@mail.mil
[2]   Monterey Bay Aquarium Research Institute, 7700 Sandholdt Rd, Moss Landing, CA 95039, USA; lkuhnz@mbari.org
[3]   Monterey Bay Aquarium, 886 Cannery Roy, Monterey, CA 93940, USA; CBiggs@mbayaq.org (C.B.); ABitondo@mbayaq.org (A.B.); MOlhasso@mbayaq.org (M.O.); mmurray@mbayaq.org (M.M.)
*   Correspondence: elise.e.ladouceur.civ@mail.mil; Tel.: +1-301-295-6196

Citation: LaDouceur, E.E.B.; Kuhnz, L.A.; Biggs, C.; Bitondo, A.; Olhasso, M.; Scott, K.L.; Murray, M. Histologic Examination of a Sea Pig (*Scotoplanes* sp.) Using Bright Field Light Microscopy. *J. Mar. Sci. Eng.* **2021**, *9*, 848. https://doi.org/10.3390/jmse9080848

Academic Editor: Azizur Rahman

Received: 22 June 2021
Accepted: 5 August 2021
Published: 6 August 2021

Publisher's Note: MDPI stays neutral with regard to jurisdictional claims in published maps and institutional affiliations.

Abstract: Sea pigs (*Scotoplanes* spp.) are deep-sea dwelling sea cucumbers of the phylum Echinodermata, class Holothuroidea, and order Elasipodida. Few reports are available on the microscopic anatomy of these deep-sea animals. This study describes the histologic findings of two, wild, male and female *Scotoplanes* sp. collected from Monterey Bay, California. Microscopic findings were similar to other holothuroids, with a few notable exceptions. Sea pigs were bilaterally symmetrical with six pairs of greatly enlarged tube feet arising from the lateral body wall and oriented ventrally for walking. Neither a rete mirabile nor respiratory tree was identified, and the large tube feet may function in respiration. Dorsal papillae protrude from the bivium and are histologically similar to tube feet with a large, muscular water vascular canal in the center. There were 10 buccal tentacles, the epidermis of which was highly folded. Only a single gonad was present in each animal; both male and female had histologic evidence of active gametogenesis. In the male, a presumed protozoal cyst was identified in the aboral intestinal mucosa, and was histologically similar to previous reports of coccidians. This work provides control histology for future investigations of sea pigs and related animals using bright field microscopy.

Keywords: holothuroidea; deep sea; elasipodida; anatomy; morphology; sea cucumbers

1. Introduction

Sea pigs (*Scotoplanes* spp.) are animals of the phylum Echinodermata, class Holothuroidea, and order Elasipodida. Echinoderms include five extant classes: Crinoidea, Asteroidea, Ophiuroidea, Echinoidea, and Holothuroidea. Echinoderms may have pseudoradial or bilateral symmetry, the latter of which is more common in holothuroids [1]. All echinoderms have a unique type of connective tissue called mutable collagenous tissue that allows them to relax and stiffen their dermis at will. Mutable collagenous tissue is necessary for evisceration and autotomy, which allows echinoderms to voluntarily expel internal viscera and release arms, respectively. Holothuroids are sea cucumbers that collectively occupy a large number of ecological niches. Approximately 1/3 of all holothuroids live in the deep sea [2].

The order Elasipodida includes holothuroids that are deep-sea dwelling [3]. The ocean covers approximately 71% of earth's surface, with about 90% of it considered deep sea (>200 m depth) [4]. The deep sea is a harsh environment with extremes in temperature, pressure, resources, and darkness [5]. Deep-sea dwellers are uncommonly examined due to the difficulty in obtaining specimens [6].

The individual specimens investigated here (*Scotoplanes* sp.) are likely representative of an undescribed species [7]. Their thus-far established depth range is from 985–1900 m, and they have been directly observed as sediment ingesters [8]. The hypothesis is that this previously unexamined, deep-sea dwelling *Scotoplanes* sp. is histologically similar to shallow-water holothuroids, which have been studied in histological detail [9–20].

2. Materials and Methods

Two sea pigs (*Scotoplanes* sp.), one male (10.8 cm total dead specimen length) and one female (8.9 cm total dead specimen length), were collected via suction sampler from the sediment-laden seafloor at 1050 m depth on Smooth Ridge, Monterey Bay, CA, USA (36°50.184′ N lat, 122°09′ W long.) using the Monterey Bay Aquarium Research Institute's *R/V Rachel Carson* and remotely operated vehicle *Ventana*. After collection, the animals collapsed and were nonresponsive as assessed by lack of body turgor and lack of any response to digital manipulation. Additional chemical sedation was not pursued as the animals were already nonresponsive, and euthanasia was accomplished through 48 h immersion in 10% neutral buffered formalin. After complete fixation, transverse, coronal sections were made of the body at 1 cm intervals and processed for histology. After initial sectioning, the dorsal papillae (also known as antennae; [20]), eviscerated digestive tract, gonads, and buccal tentacles were cut away from the body; additional transverse and longitudinal sections were made of these organs and processed for histology. All tissue sections were processed routinely for histology using a Tissue-Tek VIP 6 AI vacuum infiltration processor (Sakura Finetek, Torrance, CA, USA) using the overnight protocol. Tissues were embedded in paraffin, sectioned in 5 µm-thick sections, and stained with hematoxylin and eosin. Decalcification was not performed as there were no grossly gritty or hard tissues. Histology slides were scanned using an Aperio GT450 slide scanner (Leica Biosystems, Buffalo Grove, IL, USA), and viewed and photographed using Aperio ImageScope software (Leica Biosystems, Buffalo Grove, IL, USA).

3. Results

On gross examination, the body was bilaterally symmetrical with six pairs of large tube feet (12 feet total) extending from the trivium (i.e., ventrum) and oriented ventrally. The tube feet are the largest at mid-body and smallest near the anus. Ten specialized buccal tentacles (also called circumoral tentacles) circumferentially surround the oral cavity [21]. Two pairs of dorsal papillae are on the bivium (i.e., dorsum; Figure 1).

Histologically, the body wall consists of an epidermis and coelomic epithelium separated by the dermis. A thin cuticle covers the simple epidermis, which is composed of columnar epithelium. The body wall is variably convoluted with numerous infoldings, particularly around the buccal tentacles (Figure 2). The dermis is composed of mutable collagenous tissue and encloses canals and nerves, most of which run longitudinally along the length of the body wall. There are occasional crystalline, mineralized deposits in the dermis known as ossicles or sclerites, especially around the buccal tentacles, that polarize and frequently fall out of histologic section (artifact). The center of the animal has a large, fluid-filled coelom (also called perivisceral coelom) lined by a single layer of cuboidal to flattened coelomic epithelium.

External appendages include the tube feet, dorsal papillae, and buccal tentacles, all of which have the same histologic layers (i.e., epidermis, dermis, and myoepithelium lining a central water vascular canal). The epidermis around the tube feet contains larger numbers of secretory cells than the other appendages and is supported by subepidermal glands. Tube feet have thick longitudinal retractor muscles. Dorsal papillae and buccal tentacles are lined externally by a highly folded epidermis with an external cuticle and have a central water vascular channel similar to those seen in tube feet.

Figure 1. In situ image of *Scotoplanes* sp., taken from the remotely operated vehicle *Doc Ricketts* using an Ikegama high-definition camera fitted with a HA10Xt.2 Fujinon lens. The mouth is surrounded by 10 modified tube feet, called buccal tentacles (BT), which are used for feeding. Six pairs of tube feet (TF) line the lateral body wall and are oriented ventrally for walking. Two sets of paired elongate papillae (P) extend from the dorsal body wall (i.e., bivium). Total body length = 14 cm. Scale bar = 2 cm. MBARI 2014, 1438 m depth, Monterey Bay, CA, USA.

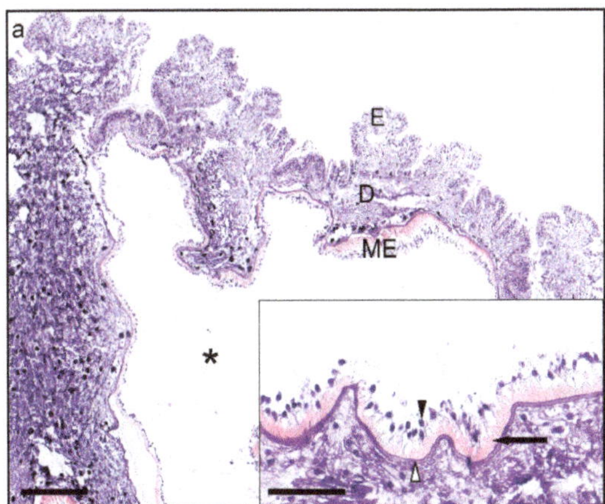

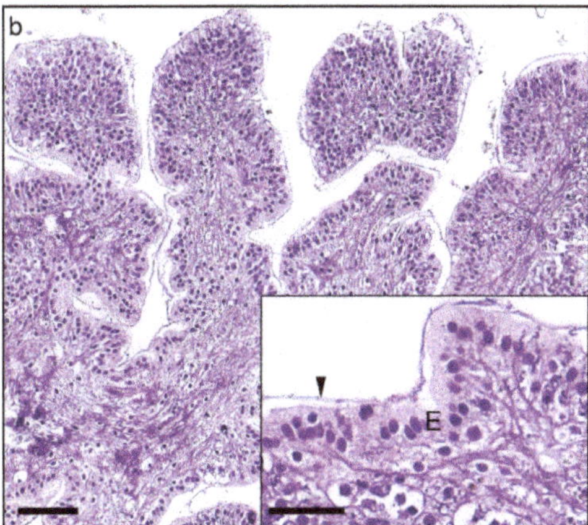

Figure 2. Histologic sections of the buccal tentacles of a sea pig (*Scotoplanes* sp.) stained with hematoxylin and eosin. (**a**) Asterisk, water vascular canal; D, dermis; E, epidermis; ME, myoepithelium. Scale bar = 200 μm. Inset: The myoepothelium has oval, apical nuclei (black arrowhead) subtended by muscle fibers (arrow) aligned on a thick basement membrane (white arrowhead). Scale bar = 60 μm. (**b**) The epidermis is highly folded and arranged into arborizing papillary projections. Scale bar = 60 μm. Inset: Arrowhead, cuticle; E, epidermis. Scale bar = 30 μm.

The alimentary canal in both animals was eviscerated prior to examination (likely due to stress of capture). Aside from the buccal tentacles and anus, the alimentary canal has a thin wall composed of a small amount of collagenous stroma with an external coelomic epithelial lining and an internal digestive mucosa. The digestive mucosa is composed of tall columnar epithelium with digestive granules oriented apically towards the lumen.

Surrounding the buccal tentacles, the anterior alimentary canal is lined internally by a cuticle and is bounded externally by muscle. Additionally, the oral mucosa is highly folded with increased numbers of digestive granules compared to aboral sections (Figure 3). In the aboral intestines of the male, a small cyst-like cavity is present in the basal region of the mucosal epithelium and filled with dozens of approximately 3 × 5 µm, oval cells with a single nucleus (presumed protozoa) (Figure 4). The lumen of the intestines contains refractile digesta.

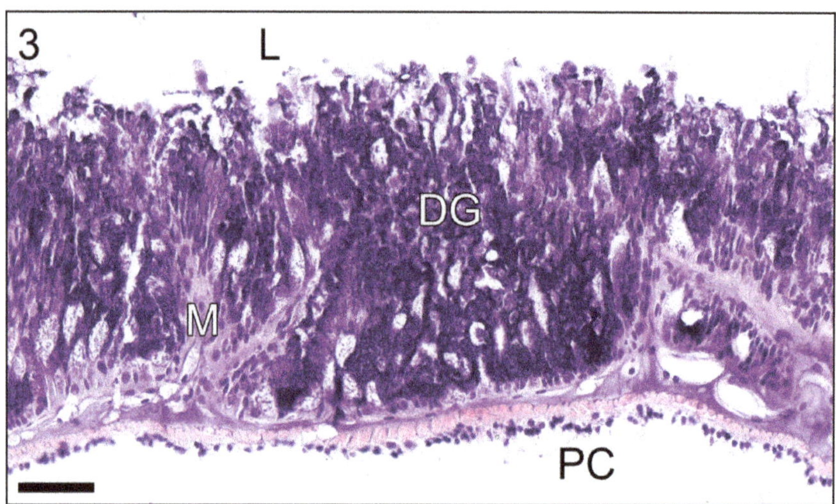

Figure 3. Histologic section of the oral mucosa of a sea pig (*Scotoplanes* sp.) stained with hematoxylin and eosin. DG, digestive granules; L, lumen; M, mucosa; PC, perivisceral coelom. Scale bar = 60 µm.

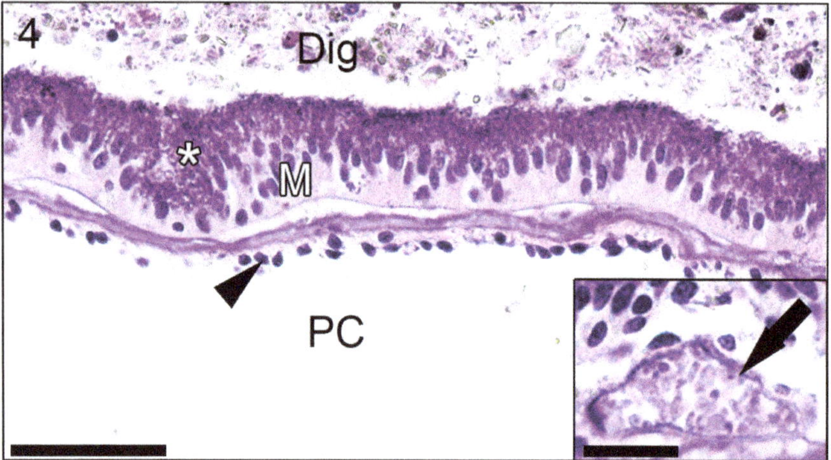

Figure 4. Histologic section of the intestines of a sea pig (*Scotoplanes* sp.) stained with hematoxylin and eosin. The lumen contains digesta (Dig). Arrowhead, coelomic epithelium; asterisk, digestive granules; M, mucosa; PC, perivisceral coelom. Scale bar = 60 µm. Inset: A cyst-like cavity (arrow) in the basilar aspect of the mucosa resembles a protozoal cyst. Scale bar = 15 µm.

Throughout the dermis and coelom, there are frequent cross-sections of water vascular canals that terminate in the tube feet. The entire water vascular system has an internal lining of cuboidal ciliated myoepithelium (called endothelium in some texts) [21,22]. The myoepithelial layer is thicker in the appendages than in the internal canals (cf. Figures 2a and 5). External to the myoepithelium is a connective tissue layer bound externally by coelomic epithelium in water vascular canals, and bound by dermis and epidermis in the tube feet [22]. The water vascular system contains clear space, coelomocytes, and some environmental debris (Figure 5). Hemal vessels consist of a thin wall of connective tissue lined externally by coelomic epithelium. Unlike the water vascular system, there is no internal lining to the hemal system. There is occasionally a small amount of muscle in the wall of the hemal vessels.

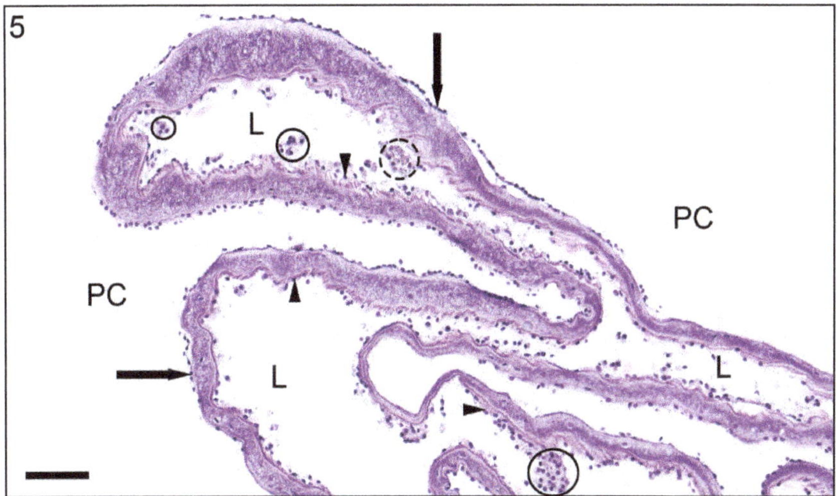

Figure 5. Histologic section of a water vascular canal of a sea pig (*Scotoplanes* sp.) stained with hematoxylin and eosin. The lumen (L) contains coelomocytes (solid circles), which are somteimes mixed with environmental debris (dashed circle). Coelomic epithelium (arrows) is occasionally artifactually detached from the canal wall. Arrowheads, myoepithelium; C, coelom; L, lumen. Scale bar = 60 μm.

Similar to other echinoderms, the nervous system consists of a network of nerves without ganglia. There is a circumoral nerve ring at the base of the buccal tentacles, and multiple branchial radial nerves that extend along the body length. Cross sections of the circumoral nerve ring are evident in cross sections of the anterior body. The nerves have peripheral cell bodies and central neuropil/axons (Figure 6).

A single gonad is present in each animal: an ovary in the female and testis in the male. The gonad consists of a thin sac of connective tissue lined externally by coelomic epithelium and lined internally by germinal epithelium. Germ cells mature centrally into previtellogenic then vitellogenic oocytes in the ovary (Figure 7). In the testis, the germ cells start as spermatogonia and mature centrally, decreasing in size to spermatocytes and finally spermatids, which are approximately 5–10 μm, spherical, and deeply basophilic (Figure 8) [23].

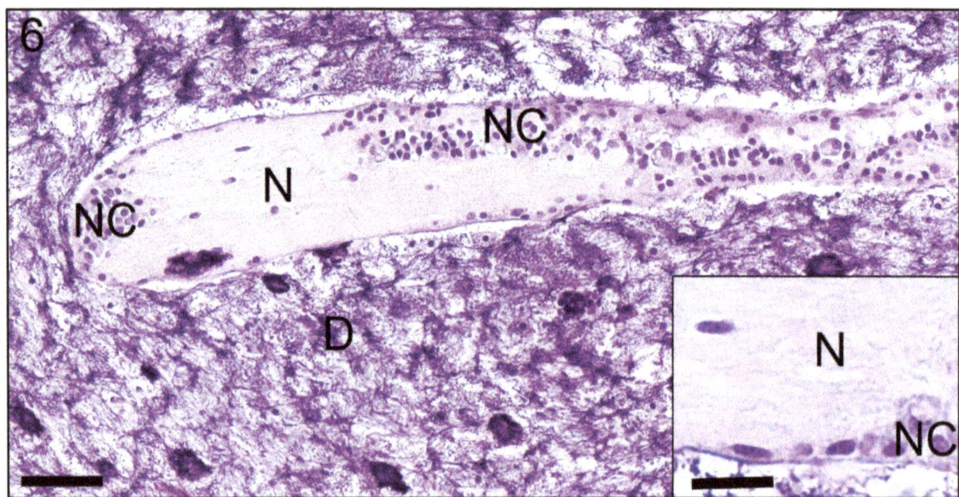

Figure 6. Histologic section of a nerve of a sea pig (*Scotoplanes* sp.) stained with hematoxylin and eosin. D, dermis; N, neuropil; NC, nerve cells. Scale bar = 60 μm. Inset: Scale bar = 30 μm.

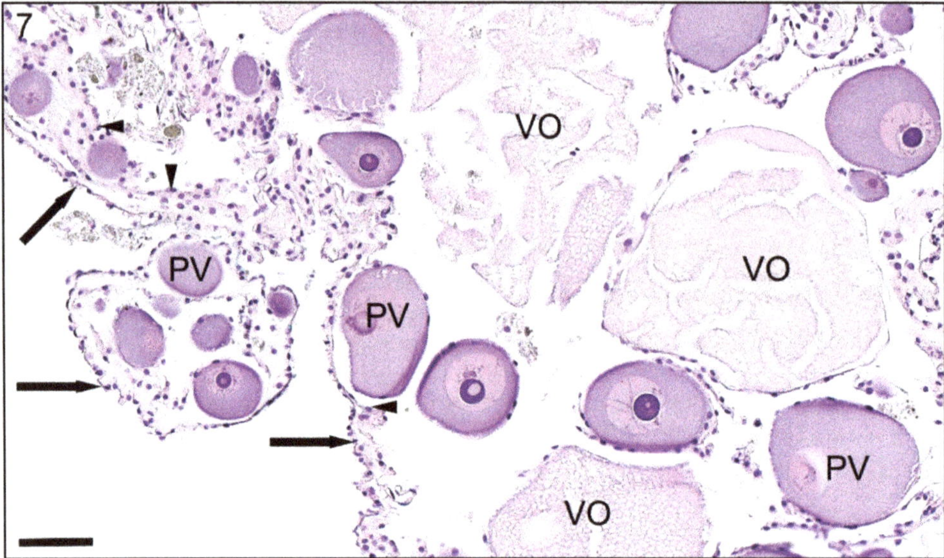

Figure 7. Histologic section of an ovary of a sea pig (*Scotoplanes* sp.) stained with hematoxylin and eosin. Vitellogenic oocytes (VO) have some artifactual loss of internal contents, creating artifactual, clear clefts. Arrows, coelomic epithelium; arrowheads, germinal epithelium; PV, previtellogenic oocyte. Scale bar = 60 μm.

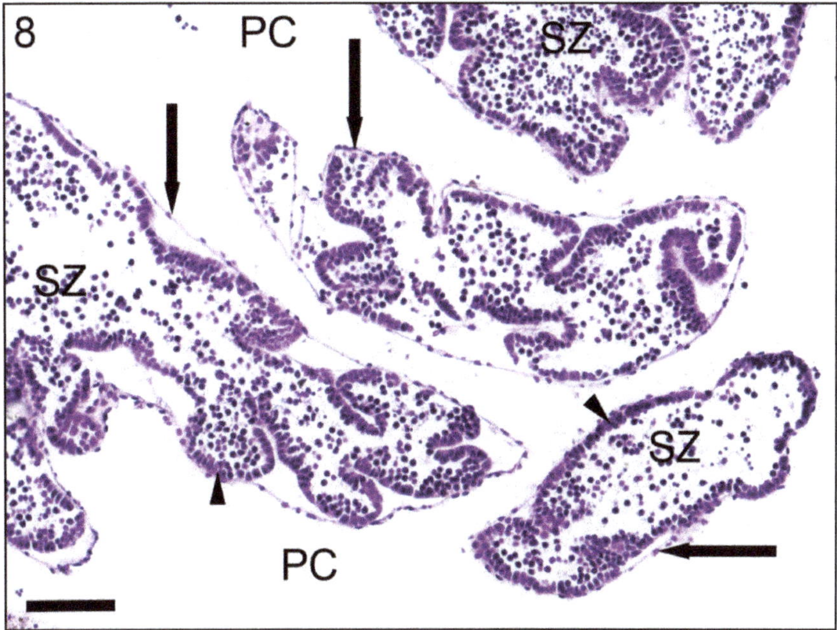

Figure 8. Histologic section of a testis of a sea pig (*Scotoplanes* sp.) stained with hematoxylin and eosin. Arrows, coelomic epithelium; arrowheads, spermatogonia; PC, perivisceral coelom; SZ, spematozoa. Scale bar = 60 μm.

4. Discussion

The results support the hypothesis that histologic findings of these deep-sea dwelling sea pigs are similar to other holothuroids, though there are few notable differences. Similarities and differences between *Scotoplanes* sp. and other holothuroids are detailed below. Ambulacral grooves are present along the polar axis of holothuroids. Depending on the species, tube feet may be concentrated in these grooves, distributed throughout the body, or be lacking altogether [2,22]. This species had markedly enlarged, bilaterally symmetrical tube feet, which are used for walking on the deep-sea ocean floor [24]. Depending on the holothurian species, the mouth may be surrounded by buccal tentacles, which are specialized tube feet that may be retracted by during periods of inanition [25]. Ten buccal tentacles were present in *Scotoplanes* sp. and had similar histologic composition to the tube feet except that the epidermis was highly folded.

In holothurians, the intestines terminate in a cuticle-lined canal that leads to the anus. This canal is the rectum in holothuroids that lack a respiratory tree, and cloaca in holothuroids that have a respiratory tree, which empties ventilator flow into the cloaca. In this *Scotoplanes* sp., a respiratory tree was not identified, which is consistent with gross anatomical reports of other Elasipodida [2]. The large tube feet of sea pigs are likely particularly important in gas exchange, which occurs through the body wall in holothuroids that lack a respiratory tree. Protozoal parasites, specifically coccidians, have been reported in some wild deep-sea elasipodids, particularly around the posterior digestive tract [24,26–28]. Structures resembling protozoa (i.e., small, elongate, uniform structures in a cyst-like cavity) were observed in the male *Scotoplanes* sp., and may represent the previously reported coccidians.

The water vascular system of holothuians is similar to other echinoderms, except the madreporite opens in the perivisceral coelom instead of in the external body wall [2,22]. The hemal and water vascular systems of this species generally appeared similar to holothuroid reports elsewhere [2,22]. Rete mirabile have been reported in some holothuroids, but were not identified in these animals, likely due to lack of a respiratory tree, with which the rete mirabile are anatomically associated [2].

Most holothurians are sexually dioecious with sexes in separate individuals. Unlike other echinoderms, holothuroids possess only a single gonad [2,15,23]. Both of these characteristics (i.e., dioecious with a single gonad) were confirmed in this *Scotoplanes* sp. Previous research suggests that in elasipodids, the testes of adult males are often inactive (lack gametogenesis), but ovaries of adult females consistently have active gametogenesis [26]. Both *Scotoplanes* sp. animals had active gametogenesis, and it is possible that this animal differs in its reproductive strategy than other elasipodids.

This study presents histologic findings in a sea pig, *Scotoplanes* sp. Microscopic studies of these and other animals in this class are limited. The information presented here may serve as controls for future studies, and provide useful information for deep-sea and holothurian taxonomists.

Author Contributions: Conceptualization, E.E.B.L., K.L.S. and M.M.; methodology, E.E.B.L., L.A.K., C.B., A.B., M.O., K.L.S. and M.M; formal analysis, E.E.B.L. and K.L.S.; investigation, E.E.B.L., L.A.K., C.B., A.B., M.O., K.L.S. and M.M.; resources, L.A.K. and M.M.; data curation, E.E.B.L., L.A.K., C.B., A.B., M.O., K.L.S. and M.M.; writing—original draft preparation, E.E.B.L., L.A.K., K.L.S. and M.M.; writing—review and editing, E.E.B.L., L.A.K., C.B., A.B., M.O., K.L.S. and M.M.; supervision, E.E.B.L. and M.M. All authors have read and agreed to the published version of the manuscript.

Funding: Collection of specimens was funded by the David and Lucile Packard Foundation via the Monterey Bay Aquarium Research Institute (901200).

Institutional Review Board Statement: The study was conducted according to the guidelines of the Joint Pathology Center, and approved by the Publication Approval Organization of Walter Reed National Military Medical Center (approval number 6217).

Informed Consent Statement: Not applicable.

Data Availability Statement: Not applicable.

Acknowledgments: Thank you also to the JPC technical and administrative support staff, including Ann Brown, Andrea Cherilus, Kenenya Gathers, Steven Mcnair, and Warren McNeil.

Conflicts of Interest: The authors declare no conflict of interest. The identification of specific products or scientific instrumentation does not constitute endorsement or implied endorsement on the part of the author, DoD, or any component agency. The views expressed in this manuscript are those of the author and do not reflect the official policy of the Department of Army/Navy/Air Force, Department of Defense, or U.S. Government.

References

1. Smirnov, A.V. Sea cucumbers symmetry (Echinodermata: Holothuroidea). *Paleontol. J.* **2014**, *48*, 1215–1236. [CrossRef]
2. Ruppert, E.E.; Fox, R.S.; Barnes, R.D. Echinodermata. In *Invertebrate Zoology*, 7th ed.; Ruppert, E.E., Fox, R.S., Barnes, R.D., Eds.; Brooks/Cole: Pacific Grove, CA, USA, 2004; pp. 872–929.
3. Théel, H. Report on Holothurioidea. Pt. I. Report of the Scientific Results of the Voyage of H.M.S. Challenger. *Zoology* **1882**, *4*, 1–176.
4. Ramirez-Llodra, E.; Brandt, A.; Danovaro, R.; De Mol, B.; Escobar, E.; German, C.R.; Levin, L.A.; Martinez Arbizu, P.; Menot, L.; Buhl-Mortensen, P.; et al. Deep diverse and definitely different: Unique attributes of the world's largest ecosystem. *Biogeosciences* **2010**, *7*, 2851–2899. [CrossRef]
5. Woolley, S.N.; Tittensor, D.P.; Dunstan, P.K.; Guillera-Arroita, G.; Lahoz-Monfort, J.J.; Wintle, B.A. Deep-sea diversity patterns are shaped by energy availability. *Nature* **2016**, *533*, 393–396. [CrossRef] [PubMed]
6. Roberts, D.; Gebruk, A.; Levin, V.; Manship, B.A.D. Feeding and digestive strategies in deposit-feeding holothurians. In *Oceanography and Marine Biology*; Barnes, M., Gibson, R.N., Eds.; Taylor and Francis: Oxford, UK, 2000; pp. 257–310.
7. Barry, J.P.; Taylor, J.R.; Kuhnz, L.A.; De Vogelaere, A.P. Symbiosis between the holothurian *Scotoplanes* sp. and the lithodid crab *Neolithodes diomedeae* on a featureless bathyal sediment plain. *Mar. Ecol.* **2016**, *38*, 1–10.
8. Jacobsen Stout, N.; Kuhnz, L.; Lundsten, B.; Schlining, K.; Schlining, B.; Von Thun, S. *The Deep-Sea Guide (DSG)*; Monterey Bay Aquarium Research Institute: Monterey, CY, USA, 2021.
9. Diaz-Balzac, C.A.; Abrea-Arbelo, J.E.; Garcia-Arraras, J.E. Neuroanatomy of the tube feet and tentacles in *Holothuria glaberrima* (Holothuroidea, Echinodermata). *Zoomorphology* **2010**, *129*, 33–43. [CrossRef]

10. Ehlers, U. Ultrastructue of the statocysts in the apodus sea cucumber *Leptosynapta inhaerens*. (Holothuroidea, Echinodermata). *Acta Zool.* **1997**, *78*, 61–68. [CrossRef]

11. Ezhova, O.V.; Ershova, N.A.; Malakhov, V.V. Microscopic anatomy of the axial complex and associated structures in the sea cucumber *Chiridota laevis* Frabricius, 1780 (Echinodermata, Holothuroidea). *Zoomorphology* **2017**, *136*, 205–217. [CrossRef]

12. Frick, J.E.; Ruppert, E.E.; Wourms, J.P. Morphology of the ovotestis of *Synaptula hydriformis* (Holothuroidea, Apoda): An evolutionary model of oogenesis and the origin of egg polarity in echinoderms. *Invertebr. Biol.* **1996**, *115*, 46–66. [CrossRef]

13. Foster, G.G.; Hodgson, A.N. Feeding, tentacle and gut morphology in five species of southern African intertidal holothuroids (Echinodermata). *S. Afr. J. Zool.* **1996**, *31*, 70–79. [CrossRef]

14. Kamenev, Y.O.; Dolmatov, I.Y.; Frolova, L.T.; Khang, N.A. The morphology of the digestive tract and respiratory organs of the holothurian *Cladolabes schmeltzii* (Holothuroidea, Dendrochirotida). *Tissue Cell* **2013**, *45*, 126–139. [CrossRef] [PubMed]

15. Kremenetskaia, A.; Ezhova, O.; Drozdov, A.L.; Rybakova, E.; Gebruk, A. On the reproduction of two deep-sea Arctic holothurians, *Elpidia heckeri* and *Kolga hyaline* (Holothuroidea: Elpidiidae). *Invertebr. Reprod. Dev.* **2020**, *64*, 33–47. [CrossRef]

16. Mashanov, V.S.; Zueva, O.R.; Heinzeller, T.; Dolmatov, I.Y. Ultrastructure of the circumoral nerve ring and the radial nerve cords in holothurians (Echinodermata). *Zoomorphology* **2006**, *125*, 27–38. [CrossRef]

17. Mashanov, V.S.; Frolova, L.T.; Dolmatov, I.Y. Structure of the digestive tube in the Holothurian *Eupentacta fraudatrix* (Holothuroidea: Dendrochirota). *Russ. J. Mar. Biol.* **2004**, *30*, 314–322. [CrossRef]

18. Utzeri, V.J.; Ribani, A.; Bovo, S.; Taurisano, V.; Calassanzio, M.; Baldo, D.; Fontanesi, L. Microscopic ossicle analyses and the complete mitochondrial genome sequence of *Holothuria (Roweothuria) polii* (Echinodermata; Holothuroidea) provide new information to support the phylogenetic positioning of this sea cucumber species. *Mar. Genom.* **2020**, *51*, 100735. [CrossRef] [PubMed]

19. Smiley, S. Holothuroidea. In *Microscopic Anatomy of Invertebrates*; Harrison, F.W., Chia, F., Eds.; Wiley-Liss: Hoboken, NJ, USA, 1994; Volume 14, pp. 401–471.

20. Hedgpeth, J.W. Galathea Report. Scientific results of the Danish deep-sea expedition round the world 1950–52. *Science* **1960**, *131*, 1521. [CrossRef]

21. Roberts, D.; Moore, H.M. Tentacular diversity in deep-sea deposit-feeding holothurians: Implications for biodiversity in the deep sea. *Biodivers. Conserv.* **1997**, *6*, 1487–1505. [CrossRef]

22. Newton, A.L.; Dennis, M.M. Echinodermata. In *Invertebrate Histology*; LaDouceur, E.E.B., Ed.; Wiley: Hoboken, NJ, USA, 2021; pp. 1–18.

23. Galley, E.A.; Tyler, P.A.; Smith, C.R.; Clarke, A. Reproductive biology of two species of holothurian from the deep-sea order Elasipoda, on the Antarctic continental shelf. *Deep Sea Res. II* **2008**, *55*, 2515–2526. [CrossRef]

24. Hansen, B. Photographic Evidence of a unique type of walking in deep-sea Holothurians. *Deep Sea Res. Oceanogr. Abstr.* **1972**, *19*, 461–462. [CrossRef]

25. Feral, J.P.; Massin, C. Digestive systems: Holothuroidea. In *Echinoderm Nutrition*; Jangoux, M., Lawrence, J.M., Eds.; CRC Press: Boca Raton, FL, USA, 1982; pp. 191–212.

26. Billett, D.S.M. The Ecology of Deep Sea Holothurians. Ph.D. Thesis, University of Southhampton, Southampton, UK, 1988.

27. Massin, C.; Jangoux, M.; Sibuet, M. Description d' Ixoreis psychropotae nov. gen., nov. sp., coccidie parasite du tube digetif de l'Holothurie abyssal Psychropotes longicauda Theel. *Protistologica* **1978**, *14*, 253–259.

28. Massin, C. Structures digestives d'holothuries Elasipoda (Echinodermata): Benthogone rosea Koehler, 1896 et Oneirophanta mutabilis Theel, 1879. *Arch. Biol.* **1984**, *95*, 153–185.

Journal of
Marine Science and Engineering

MDPI

Article

Intestine Explants in Organ Culture: A Tool to Broaden the Regenerative Studies in Echinoderms

Samir A. Bello *,†,‡ and José E. García-Arrarás *

Department of Biology, University of Puerto Rico, Rio Piedras Campus, P.O. Box 23360, San Juan, PR 00931, USA
* Correspondence: samir.bello@upr.edu (S.A.B.); jose.garcia36@upr.edu (J.E.G.-A.)
† Current affiliation: Department of Chemistry, University of Puerto Rico, Rio Piedras Campus, P.O. Box 23346, San Juan, PR 00931, USA.
‡ Current affiliation: Molecular Sciences Research Center, University of Puerto Rico, 1390 Ponce De León Ave, Suite 1-7, San Juan, PR 00926, USA.

Abstract: The cellular events underlying intestine regrowth in the sea cucumber *Holothuria glaberrima* have been described by our group. Currently, the molecular and signaling mechanisms involved in this process are being explored. One of the limitations to our investigations has been the absence of suitable cell culture methodologies, required to advance the regeneration studies. An *in vitro* system, where regenerating intestine explants can be studied in organ culture, was established previously by our group. However, a detailed description of the histological properties of the cultured gut explants was lacking. Here, we used immunocytochemical techniques to study the potential effects of the culture conditions on the histological characteristics of explants, comparing them to the features observed during gut regeneration in our model *in vivo*. Additionally, the explant outgrowths were morphologically described by phase-contrast microscopy and SEM. Remarkably, intestine explants retain most of their original histoarchitecture for up to 10 days, with few changes as culture time increases. The most evident effects of the culture conditions on explants over culture time were the reduction in the proliferative rate, the loss of the polarity in the localization of proliferating cells, and the appearance of a subpopulation of putative spherulocytes. Finally, cells that migrated from the gut explants could form net-like monolayers, firmly attached to the culture substrate. Overall, regenerating explants in organ culture represent a powerful tool to perform short-term studies of processes associated with gut regeneration in *H. glaberrima* under controlled conditions.

Keywords: sea cucumber; intestine explant; *in vitro*; regeneration; organ culture

Citation: Bello, S.A.; García-Arrarás, J.E. Intestine Explants in Organ Culture: A Tool to Broaden the Regenerative Studies in Echinoderms. *J. Mar. Sci. Eng.* **2022**, *10*, 244. https://doi.org/10.3390/jmse10020244

Academic Editor: Alexandre Lobo-da-Cunha

Received: 31 December 2021
Accepted: 7 February 2022
Published: 11 February 2022

Publisher's Note: MDPI stays neutral with regard to jurisdictional claims in published maps and institutional affiliations.

1. Introduction

Explants are non-disaggregated tissue or organ fragments removed from an organism. In mammals, cultured explants are generally used to obtain primary cell lines constituted by the cells that migrate out from the tiny explants as outgrowths, forming monolayers around them. Remarkably, in most cases, the original pieces of tissue or explants are discarded [1]. Two variants of this method have been developed to study the tissues themselves, rather than the isolated cells: organotypic cultures, and explants in organ culture. In organotypic cultures, cells of different lineages from a donor organ are isolated, expanded, and grown in three-dimensional scaffolds to generate tissue equivalents. Explants in organ culture, in contrast, refer to non-disaggregated tissues or whole small organs (such as mice kidneys) maintained *in vitro*, retaining most of the *in vivo* tissue architecture [1]. Organ culture is the preferred method to study histological changes or the behavior of tissues in response to external stimuli, rather than the response of individual cells [2]. Explants in organ culture and their outgrowths have been studied in echinoderms, mainly in the classes Crinoidea and Ophiuroidea [3–7].

The arm explant model has been the most used in echinoderms for *in vitro* studies. Arm explants have been utilized in organ culture to study regenerative processes, mainly using histological methods, while some groups have focused on describing their outgrowths. Pioneering studies by Candia-Carnevali et al. [3] demonstrated that the cellular events observed in regenerating arms of the crinoid *Antedon mediterranea in vivo* also occur in arm explants, which can survive *in vitro* for several weeks. Similar findings were obtained by Dupont and Thorndike [4] and Burns et al. [5] using double amputated arm explants from the brittle star *Amphiura filiformis*. Additionally, the outgrowths obtained around arm explants of the species *A. mediterranea* were studied by Di Benedetto et al. [6]. Few explants from other echinoderm organs have been cultured. Moss et al. [7] used radial nerve cord and oral plate explants from the sea star *Asterias rubens* and the brittle star *Ophiura ophiura*, respectively, to study the cells that were able to migrate from them. Recently, our group established an *in vitro* system using explants in the class Holothuroidea [8].

Adult holothurians are a powerful model for organ regeneration, due to their fantastic ability to spontaneously regrow complex organs after autotomy [9]. Many of the cellular and molecular processes underlying the formation of a new intestine from the mesentery after evisceration in the species *Holothuria glaberrima* have been described [10]. Several techniques have been employed in *in vivo* studies to decipher these processes; however, crucial aspects, such as the genomic and signaling regulation of the organ regrowth, remain to be elucidated. The lack of *in vitro* systems in holothurians has hampered the implementation of new techniques to advance the understanding of organogenesis under controlled conditions, free of systemic variations that might arise *in vivo*. Thus, we have established the explant culture, which was recently used to gain new insights about the signaling mechanisms involved in intestine regrowth [8,11] and the dedifferentiation mechanisms during radial nerve cord regeneration [12] in our model. Furthermore, intestine explants in organ culture were crucial to the establishment of gene knockdown by the RNA interference in *H. glaberrima* [13].

Here, we implemented the methodology to maintain and characterize regenerating intestines of *H. glaberrima* cultured for 24 h, 5 d, or 10 d. For this, we used immunocytochemical techniques (antibodies both commercially available and produced by our group). We emphasized the effects of the culture conditions on cellular processes that have been well documented during gut regeneration *in vivo* in our model, such as muscle dedifferentiation, cell proliferation, changes in the extracellular matrix components, and the distribution of cell populations, such as neurons and mesothelial cells in explants. Finally, the explant outgrowth characteristics were described by phase-contrast microscopy and SEM.

2. Materials and Methods

2.1. Animal Collection and Evisceration

Adult sea cucumbers from the species *Holothuria glaberrima* were collected from the rocky shores of northeastern Puerto Rico. Upon arrival to the laboratory, animals were injected with 3–5 mL of 0.35 M KCl into the coelomic cavity to induce evisceration. After evisceration, they were maintained in indoor seawater aquaria at room temperature (RT, 20–24 °C) with constant aeration until sacrificed. Animals were anesthetized by immersion in 0.2% 1,1,1-trichloro-2-methyl-2-propanol hemihydrate (chlorobutanol, Sigma-Aldrich, St. Louis, MO, USA) in natural sea water for 20–30 min before being sacrificed.

2.2. Explants

Gut rudiments and associated mesentery were dissected under a stereoscopic microscope from sea cucumbers undergoing intestine regeneration at 5 days post-evisceration (dpe). Immediately after dissection, gut rudiment explants were surface disinfected, as described previously [14], and carefully placed in 24-well plates, one explant per well. Subsequently, each explant was covered with 1 mL of L-15 culture medium (Leibovitz, Cat. L4386, Sigma-Aldrich, St. Louis, MO, USA) conditioned for marine species adding salts as described [14,15] and adjusted to pH = 7.8. The culture medium was also supplemented with antibiotics (100 U/mL penicillin, 100 µg/mL streptomycin, 50 µg/mL gentamicin), antifungal (2.5 µg/mL amphotericin B), 1X MEM nonessential amino acids, 1 mM sodium pyruvate, 1.75 µg/mL α-tocopherol acetate, and 1X ITS (Insulin, Transferrin, and Sodium Selenite). All cell culture reagents were purchased from Sigma-Aldrich (St. Louis, MO, USA). Explants were incubated in a modular incubator chamber (Billups-Rothenberg Inc., San Diego, CA, USA) at room temperature (20–24 °C) for 24 h, 5 d, or 10 d. Culture medium was changed every other day. Control explants were 5 dpe regenerating intestines from sea cucumbers that were dissected and disinfected as the experimental ones, but that were fixed without placing in culture.

2.3. Characterization of Cell Populations in Cultured Explants by Immunocytochemistry

After incubation, explants were fixed in 4% PFA diluted in 0.1 M PBS at 4 °C overnight (ON). After fixation, tissues were rinsed three times with PBS (15 min each) and cryoprotected in 30% sucrose/PBS at 4 °C. Intestine explants were mounted in OCT compound (Sakura Finetek Inc, Torrence, CA, USA) and cryosectioned (20 µm) using a cryostat (model CM1850, Leica Microsystems, Nussloch, Germany). Immunohistochemical staining of tissue sections was performed, as previously described [16]. Briefly, tissue sections were permeabilized with 1% Triton X-100 for 15 min, followed by two rinses with PBS. Then the tissues were blocked with normal goat serum, and subsequently they were incubated ON at room temperature in a humid chamber with the primary antibodies shown in Table 1. These antibodies have been widely used to identify different cell populations in holothurian tissues in *in vivo* studies [16–22]. Next day, the slides were washed three times with PBS for 15 min each wash. Then, the tissue sections were incubated with the secondary antibody goat anti-mouse labeled with Cy3 (GAM-Cy3, 1:1000) or goat anti-rabbit labeled with Cy3 (GAR-Cy3, 1:1000) from Jackson ImmunoResearch Labs (West Grove, PA, USA) for an hour at room temperature. The same methodology was used for double immunolabeling, except that tissue sections were incubated simultaneously with both primary antibodies, and the next day with both secondary antibodies, GAM-Cy3 (1:1000), and GAR-Alexa Fluor 488 (Thermo Fisher Scientific, Rockford, IL, USA). After three additional washes with PBS, slides were mounted in buffered glycerol solution containing 1 µg/mL of 4′,6-diamidino-2-phenylindole (DAPI, Sigma-Aldrich, St. Louis, MO, USA). Finally, tissue sections were visualized and analyzed using a fluorescence microscope (Eclipse Ni, Nikon Instruments Inc., Melville, NY, USA).

Table 1. Primary antibodies used for immunohistochemical studies.

Antigen (Antibody Name)	Raised in	Immunogen	Cells/Structures Labeled	Source	Dilution
Unknown (Meso-1)	Mouse (monoclonal)	Homogenate of the regenerating mesothelium	Peritoneocytes and myocytes of the mesothelium	Dr. García-Arrarás lab. [16]	1:50
β-Tubulin (Anti- β-Tubulin clone	Mouse (monoclonal)	Tubulin from *S. purpuratus* sperm axonemes	Subpopulation of neuron-like cells	Sigma-Aldrich (T-293) [17]	1:500
HgSTARD10 protein (RN1)	Mouse (monoclonal)	Homogenate of the radial nerve of *H. glaberrima*	Subpopulation of neuron-like cells	Dr. García-Arrarás lab. [18,19]	1:50,000
GFSKLYFamide (GFS)	Rabbit (polyclonal)	Synthetic peptide GFSKLYFamide	Subpopulation of neuron-like cells	Dr. García-Arrarás lab. [20]	1:1000
Unknown (Sph3)	Mouse (monoclonal)	Holothurian homeobox peptide coupled to bovine serum albumin	Spherulocytes	Dr. García-Arrarás lab. [21]	Undiluted
Fibrous collagen (Hg-fCOL)	Mouse (monoclonal)	Insoluble intestinal fibrous collagen extract of *H. glaberrima*	Fibrous collagen	Dr. García-Arrarás lab. [22]	1:10
Bromodeoxyuridine (Anti-BrdU)	Mouse (monoclonal)	BrdU	Cell nuclei of replicating cells	Millipore-Sigma (GERPN202)	1:5

2.4. Gut Rudiment Area

Following evisceration, the mesentery is attached to the body wall, which comprises its proximal boundary. At the other end it is 'free' within the coelomic cavity. As the 'free' end of the mesentery forms the new intestinal rudiment, a constriction forms that separates the growing intestine from the remaining mesentery. Both boundaries are clearly observed in the explant tissue sections. Thus, we took photographs of gut rudiments in DAPI-labeled tissue sections of control and cultured explants (24 h, 5 d, or 10 d) using a Nikon DS-Q12 digital camera attached to the fluorescence microscope. Gut rudiment area was measured on photographs using the freely available image processing program Image J (version Fiji 1.46, NIH, MD, USA) downloaded from https://imagej.nih.gov/ij/ and last accessed on 15 December 2021. At least three non-consecutive tissue sections were evaluated to obtain the average of gut rudiment area per explant. At least 4 explants ($n = 4$–5) were used for each culture time point and control.

2.5. Muscle and Spindle-like Structures (SLSs)/DAPI Ratio in Cultured Explants

Muscle and spindle-like structures were detected in tissue sections using Phalloidin-TRITC staining (1:1000, Cat. P1951, Sigma-Aldrich, St. Louis, MO, USA). Slides were incubated with diluted Phalloidin-TRITC for one hour, then washed three times with PBS and mounted in buffered glycerol solution containing DAPI. Finally, tissue sections were visualized and analyzed using a Nikon Eclipse Ni fluorescence microscope.

The number of SLSs and DAPI labeled nuclei per field of view were manually counted under the fluorescence microscope, using the 100X objective (area = 13,837 μm^2) for each tissue section on three different regions of the mesentery: near the gut rudiment (distal mesentery), medial segment (medial mesentery), and near the body wall (proximal mesentery) in explants cultured for 24 h, 5 d, or 10 d. The number of SLSs and cell nuclei in the three regions was averaged to obtain the SLSs/DAPI ratio per tissue section. At least two non-consecutive tissue sections were used per explant. A total of 5 explants were used in this experiment for each time point ($n = 5$).

2.6. Cell Proliferation Assays

Cell proliferation was evaluated in explants from 5 dpe after 24 h, 5 or 10 days in *in vitro* conditions by BrdU incorporation assay. BrdU (Sigma-Aldrich, St. Louis, MO, USA) was added to the cell culture medium at a final concentration of 50 μM for the last 24 h of culture. For explants maintained for only 24 h *in vitro*, we added the BrdU at the initial explant seeding. Detection of BrdU labeling was performed by immunocytochemistry. The processing of explant tissue sections for anti-BrdU staining was similar to that used for other markers; however, it included an additional step to denature the DNA after fixation. DNA denaturation was done by treating tissue sections with 0.05 M HCl for 1 h after permeabilization. After DNA denaturation, tissue sections were washed twice with PBS, for 10 min each, and then blocked with normal goat serum (diluted 1:50) for 1 h at RT. Subsequently, cells were incubated with the primary antibody anti-BrdU (1:5, Cat. GERPN202, Millipore-Sigma, Marlborough, MA, USA) ON. Next day, tissue sections were washed three times with PBS for 15 min each. Then, tissue sections were incubated with the secondary antibody GAM-Cy3 (1:1000), mounted, and visualized as was mentioned previously.

The percentage of BrdU-positive cells was calculated in at least two non-consecutive tissue sections of gut explants after 24 h, 5 d, or 10 d of culture. For each culture time, 8–10 explants ($n = 8$–10) were used. To obtain the percentage of BrdU positive cells, photographs were taken of both the gut rudiment and the mesentery (in three regions: distal, medial, and proximal mesentery) with 40X objective (area = 85,580 μm^2) and imported to Image J software. The percentage of BrdU positive cells in each region was calculated and averaged to obtain the percentage of BrdU positive cells per explant. DAPI is a nuclear dye that labels all nuclei; thus, providing the number of total cells within the field of view, so as to obtain the percentage of total cells labeled by the BrdU.

2.7. Morphological Characterization of Explant's Outgrowths by SEM

Explants from 5 dpe animals were seeded on glass cover slips of 8 mm in diameter (Electron Microscopy Sciences, Hatfield, PA, USA) coated with poly-L-lysine and cultured for 24 h, 5 d, or 10 d. After incubation, the cells that were able to migrate from explants and that remained adhered to the glass coverslips were fixed and dehydrated, as described in [23], and chemically dried and mounted, as described in [14] with a few modifications. Briefly, cells were fixed in 4% glutaraldehyde in natural sea water instead of PBS.

2.8. Statistical Analysis

To evaluate the statistical differences in gut rudiment area, SLS/DAPI ratio, and proliferation rate between control and cultured explants, we employed one-way ANOVA followed by Tukey's test for multiple comparisons of means. All values are reported as mean ± standard deviation (SD). The differences were considered significant when the p-value was <0.05. The level of significance was denoted in figures with asterisks as follow: **** $p < 0.0001$, *** $p < 0.001$, ** $p < 0.01$, * $p < 0.05$, ns = not significant (p-value > 0.05). These analyses were performed using the software GraphPad Prism 8 (San Diego, CA, USA).

3. Results

Explants from 5 dpe were cultured for 24 h, 5 d, or 10 d. The intestine rudiment area and the expression and localization of markers for muscle dedifferentiation, cell proliferation, mesothelium, nerve fibers and neuron-like cells, and components of the connective tissue were determined in the above-mentioned cultured gut tissues and were compared to explants dissected at 5 dpe but not cultured (controls).

3.1. Gut Rudiment Area

The gut rudiment grows as intestine regeneration proceeds in *H. glaberrima in vivo* [16]. Here we determined if the intestine rudiment in explants increases in size as the time in culture is extended, mimicking the growth observed *in vivo*. Remarkably, no significant increase in the gut rudiment area was observed as the culture time increased (Figure 1A–D). Control and cultured explants showed gut rudiment areas in the range of 2 to 3×10^5 μm^2 on average (Figure 1E), which agrees with values reported for intestine rudiments at 5 dpe in our model *in vivo* [16]. Overall, our results suggest that explants did not change their size as the time in culture advanced.

3.2. Muscle Dedifferentiation

Muscle cells, originally assembled into bundles in the mesentery, undergo a process of dedifferentiation during intestinal regeneration. Muscle disassembly and the concomitant appearance of spindle-like-structures (SLSs) are a hallmark of muscle dedifferentiation in echinoderms [3,24,25] and occur following a temporal and spatial sequence [9,16]. Here, we found that in control explants and in those cultured for 24 h, 5 d, and 10 d the muscle fibers had disappeared from the gut rudiment and adjacent mesentery (distal mesentery) but remained from the medial to the proximal mesentery (Figure 2A–D). However, explants cultured for 5 and 10 days exhibited disorganized muscle fibers in the proximal mesentery (Figure 2G,H) when compared to control explants and explants cultured for 24 h (Figure 2E,F). Regarding SLSs, they had disappeared from the gut rudiment but were still present in the mesentery of control and cultured explants (Figure 2A–D). The same characteristics were observed in sea cucumbers regenerating their intestines at 5 dpe (*in vivo* conditions) [16] (as shown in Figure A1), except for the muscle fiber disorganization observed in explants cultured for 5 and 10 days.

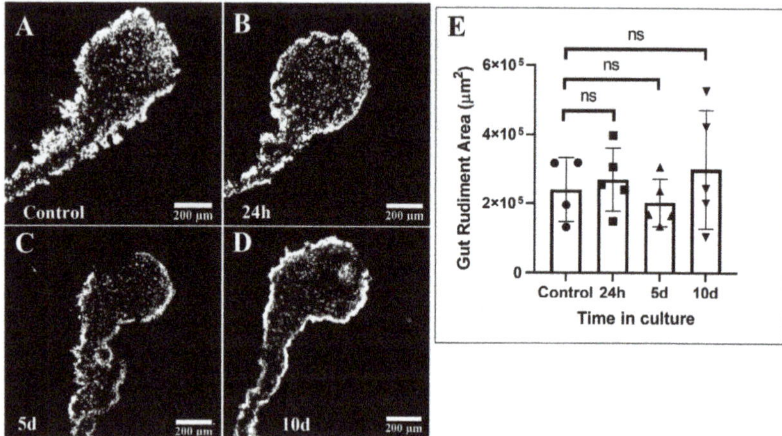

Figure 1. Gut rudiment area in cultured explants. Representative micrographs of gut rudiments in control (5 dpe, not cultured) (**A**), and explants cultured for 24 h (**B**), 5 d (**C**), and 10 d (**D**) are shown. Gut rudiment area was measured in photographs using ImageJ software. No significant differences in the gut rudiment area were found between control and cultured explants (**E**). Columns represent the mean ± SD for each group of explants, and circles, squares, triangles are the individual values of each explant. One-way ANOVA. ns = not significant differences (*p*-value > 0.05), *n* = 4 for controls, and *n* = 5 for cultured explants. Cell nuclei (DAPI) are shown in white. Scale bar = 200 μm.

The SLS/DAPI ratio can be used to indicate the degree of muscle dedifferentiation in *H. glaberrima*, where a higher ratio would suggest increased muscle dedifferentiation [8]. As mentioned above, muscle differentiation occurs in a spatial sequence in the mesentery. Thus, we determined the SLS/DAPI ratio in the three different regions of the mesentery: distal, medial, and proximal, to verify potential changes at different mesenterial locations. We observed a trend toward a SLS/DAPI ratio decrease at the distal mesentery in cultured explants with increased time in culture; however, the reduction was significant only between explants cultured for 24 h compared to explants cultured for 10 days. Conversely, we did not observe significant differences in the SLS/DAPI ratio at the medial mesentery; whereas, at the proximal region, a significant increase in the ratio was observed between control and explants cultured for 5 d. When the SLS/DAPI ratios of the three mesenteric regions were averaged for each sample, no significant differences were found between control and cultured explants, suggesting that *in vitro* conditions, per se, did not induce muscle dedifferentiation.

3.3. Cell Proliferation

The ability of cells to proliferate at early regenerative stages is an essential source of new cells for intestine regrowth. Here we evaluated the ability of cells in explants cultured for 24 h, 5 d, or 10 d to proliferate using a BrdU incorporation assay. The percentage of BrdU positive cells was determined in each region of the explants: gut rudiment and distal, medial, and proximal mesentery, to verify if the pattern in the localization of BrdU positive cells was similar in control and cultured explants. To confirm the localization of the gut rudiment and the proximal mesentery (which can be confounded) in cultured explants, we labeled tissue sections of the same explants with phalloidin (as the polarity in the localization of SLS is maintained, as was shown in Section 3.2) and anti-BrdU (See Figure A2). We found that the percentage and distribution pattern of BrdU immunoreactive cells differed in explants, according to the region analyzed and the time in culture (Figure 3). In explants cultured for 24 h, a higher percentage of BrdU positive cells were observed at the gut rudiment and distal mesentery (16.2 ± 9% and 15.0 ± 6.8%, respectively) (Figure 3A,D,G,H) compared to the medial and proximal mesenteric regions (5.5 ± 4% and

4.7 ± 5.1%, respectively) (Figure 3A,D,I,J). In explants cultured for 5 d, the percentage of proliferating cells was relatively constant (in the range of 7 to 9%) in the different regions of the explants (Figure 3B,E,G–J). On the contrary, in explants cultured for 10 d, a smaller percentage of BrdU positive cells was observed in the gut rudiment (1.03 ± 0.7%), and this value increased in the distal (2.6 ± 1.4%), medial (3.5 ± 3.1%), and proximal (3.7 ± 3.1%) regions of the mesentery (Figure 3C,F,G,J). In gut rudiments at 5 dpe in *in vivo* conditions, the BrdU positive cells were mainly observed in the gut rudiment and in the mesenterial region close to it (a portion of the distal mesentery) [16], similarly to the pattern observed here in explants cultured for 24 h.

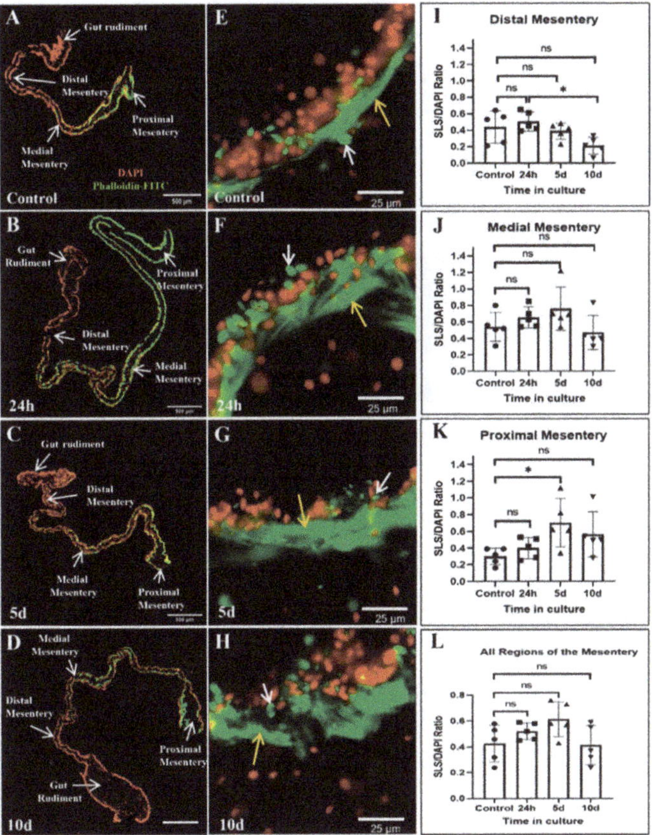

Figure 2. Spindle-like structures (SLSs) and muscle fiber distribution in cultured gut explants. The presence and distribution of SLSs and muscle fibers in control (not cultured) (**A**) and cultured for 24 h (**B**), 5 d (**C**), and 10 d (**D**) explants are shown. Tissue sections of the proximal mesentery in control (**E**), and cultured for 24 h (**F**), 5 d (**G**), and 10 d (**H**) are displayed, showing muscle bundle disorganization in explants cultured for 5 and 10 d. The quantification of the SLS/DAPI ratio in distal (**I**), medial (**J**), and proximal (**K**) mesentery is shown. SLS/DAPI ratio of all regions of the mesentery is displayed in (**L**). When the SLSs/DAPI ratios in all regions of explants are considered, no significant differences between control and cultured explants are observed (**L**). Columns represent the mean ± SD for each group of explants, and circles, squares, triangles are the individual values of each explant. One-way ANOVA, * *p* < 0.05, ns = not significant differences (*p*-value > 0.05). *n* = 5 in all groups. All regions (gut rudiment and distal, medial, and proximal mesentery) are labeled in each explant in (**A–D**). White arrows point to SLSs and yellow arrows to muscle fibers in (**E–H**). SLSs and muscle fibers are shown in green and cell nuclei (DAPI) in red. Scale bar = 500 μm in (**A–D**) and 25 μm in (**E–H**).

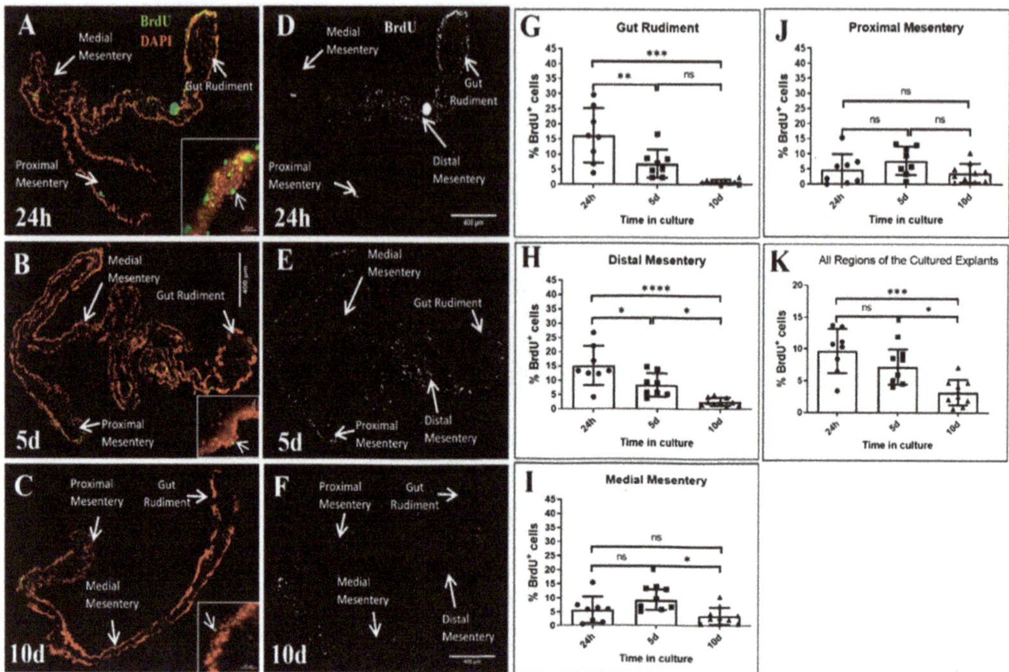

Figure 3. Cell proliferation assessed by BrdU incorporation in cultured explants. BrdU positive cell nuclei were observed in explants cultured for 24 h (**A,D**), 5 d (**B,E**), and 10 d (**C,F**). The percentage of BrdU positive cells in gut rudiment (**G**), and distal (**H**), medial (**I**), and proximal (**J**) regions of the mesentery in explants are shown. The percentage of BrdU positive cells averaging the counts performed in all regions of each explant are shown in (**K**). When the percentage of BrdU positive cells in all regions of explants are considered, a significant reduction in the number of BrdU positive cells was observed in explants cultured for 10 d compared to explants cultured for 24 h and 5 d (**G**). Columns represent the mean ± SD for each group of explants, and circles, squares, and triangles are the individual values of each explant. One-way ANOVA, **** $p < 0.0001$, *** $p < 0.001$, ** $p < 0.01$, * $p < 0.05$, ns = not significant differences (p-value > 0.05). $n = 8$–10. In Panels (**A–C**) all cell nuclei are shown in red (DAPI), with BrdU positive nuclei (white arrows in inset in **A–C**) in green. In panels (**D–F**) the BrdU positive nuclei are shown in white. Bar = 400 μm.

These findings suggest that the polarity in the localization of proliferating cells changes in explants as the time in culture increases. When the percentage in BrdU cells found in the different regions of the explants were averaged for each sample, a significant reduction in the percentage of BrdU-labeled cells in explants cultured for 10 days compared with explants cultured for 24 h and 5 days (a reduction of 67% and 45%, respectively) was observed (Figure 3K). In explants cultured for 5 days there was a trend toward proliferation rate reduction, but no significant difference was found when compared to 24 h (Figure 3K). Overall, our results suggest changes in both the proliferation rate and the polarity in proliferating cells' localization in explants cultured for 5 or more days.

3.4. Mesothelium Labeling in Cultured Explants

The monoclonal antibody (Meso-1), developed in our lab, labels the intestinal and mesenteric mesothelial cells [16]. Here, we observed that control and cultured explants showed a similar pattern of intestinal and mesenteric mesothelium labeling with Meso-1 as that observed in 5 dpe animals. Meso-1 immunoreactive cells ingress from the mesothelium to the connective tissue in control and in explants cultured for 24 h (Figure 4B,F yellow arrows), and it appears to decrease after 5 and 10 days of culture (Figure 4J,N yellow arrows). Overall, culture of explants did not cause major changes in the intestinal or mesenteric mesothelium at the histological level.

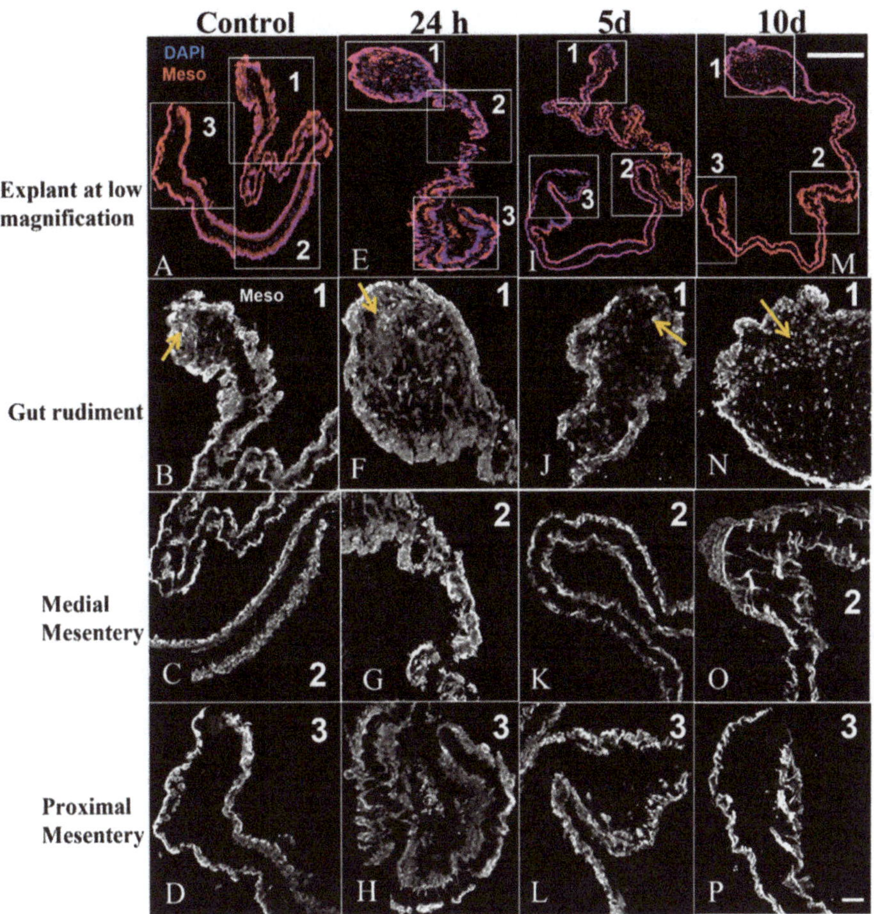

Figure 4. Meso-1 staining pattern in cultured explants from 5 dpe. Control (**A**) and explants cultured for 24 h (**E**), 5 d (**I**), and 10 d (**M**) at low magnification are shown in the upper row. The boxed areas labeled as 1, 2, and 3 in these photographs correspond to gut rudiment, medial, and proximal mesenteries, respectively, shown at a higher magnification from the second to the fourth rows. Meso-1 immunoreactive mesothelium in gut rudiment of control (**B**), and explants cultured for 24 h (**F**), 5 d (**J**), and 10 d (**N**) are shown in the second row, while immunoreactive mesothelium in the medial mesentery of control (**C**), and explants cultured for 24 h (**G**), 5 d (**K**), and 10 d (**O**) are shown in the third row. Meso-1 immunoreactive mesothelium in the proximal mesentery of control (**D**) and explants cultured for 24 h (**H**), 5 d (**L**), and 10 d (**P**) are shown in the fourth row. Meso-1 immunoreactive mesothelium is shown in red and cell nuclei labeled with DAPI are shown in blue. Bar = (**A,E,I,M**) 500 μm, (**B–D,F–H,J–L,N–P**) 100 μm.

3.5. Nerve Fibers and Neuron-like Cells Labeled by Neuronal Markers in Cultured Explants

RN1. The monoclonal antibody RN1 was developed in our laboratory as a neuronal marker for holothurian tissues [18]. As has been shown in previous results of 5 dpe animals, in our controls and in explants cultured for 24 h, the RN1 antibody labeled disorganized nerve fibers, mainly in the connective tissue, but no neuron-like cells in the gut rudiment (Figure 5A,B,E,F). In the mesentery, both nerve fibers and neuron-like cells were observed as associated with the mesenterial mesothelium (mesothelial plexus) and in the connective tissue (connective tissue plexus) (Figure 5C,D,G,H). The RN1 immunoreactive cells in the mesenteric connective tissue were mainly ovoid-shaped with a short cell projection (Figure 5D, inset). In explants cultured for 5 or 10 d, only remnants of nerve fibers were observed at the gut rudiment (Figure 5I,J,M,N), while in the mesentery, nerve fibers and neuron-like cells were observed associated with the mesothelium, and only fibers were seen in the connective tissue (Figure 5K,L,O,P). Interestingly, a smaller number of nerve fibers appeared to be present in cultured explants over time. In summary, nerve fibers and neuron-like cells immunoreactive for RN1 were observed in both control and cultured explants, but a reduction in their number was found with culture time.

Anti-β-Tubulin. This antibody is a neuronal marker that has been used to label both neural cells and fibers in *H. glaberrima in vivo*; however, it may also recognize other cell types, such as peritoneocytes and non-neural connective tissue cells. The labeling pattern varies depending on the antibody used, its dilution, and the regenerative stage studied [17]. Here, we observed a similar staining pattern, with the anti-β-tubulin antibody in control and explants cultured for 24 h, 5 d, and 10 d. Mesothelial cells or peritoneocytes, from both the gut rudiment and the mesentery were immunoreactive for anti-β-tubulin. Although the boundaries of peritoneocytes were not clear, in some areas it was evident that they were polygonal or columnar (Figure 6B,G,L,Q). Additionally, some of the cells of the connective tissue were also labeled with this antibody. Most of these cells were spherical, but some were elongated or irregular in shape (Figure 6C,H,M,R). Nerve fibers immunoreactive for β-tubulin were observed mainly in the proximal mesentery, although scarce fibers were also seen in the medial mesentery (Figure 6D,E,I,J,N,O,S,T). No immunoreactive fibers to β-tubulin were observed at the gut rudiment or distal mesentery (Figure 6B,G,L,Q). Overall, a similar staining pattern of neural and non-neural components in control and cultured explants by anti- β-tubulin was observed here.

GFS. The polyclonal antibody developed against the neuropeptide GFSKLYFamide (hereafter referred as anti-GFS) has been used in our model *H. glaberrima* as a neuronal marker in normal and regenerating tissues [17,20,26]; however, it may also recognize nonspecifically spherulocytes or morula cells [26]. Here, we observed differences in the tissue components labeled with anti-GFS between control and cultured explants. In control explants, we observed GFS immunoreactive nerve fibers mainly in the connective tissue of the gut rudiment or associated with the mesenterial mesothelium (mesothelial plexus) (Figure 7A–D). No cells immunoreactive for GFS were observed in control explants in gut rudiment or mesentery. A similar pattern of GFS labeling was observed in explants cultured for 24 h; however, few GFS-immunoreactive ovoid, or irregular, shaped cells were observed in the connective tissue of the proximal mesentery. These cells somewhat resembled morula cells or spherulocytes and were never observed to be labeled with anti-GFS in tissues of control animals. Additionally, a smaller number of nerve fibers were seen in gut rudiment and mesentery (Figure 7E–H). Some of the mesenterial mesothelial cells (medial and proximal regions) were immunoreactive for GFS after 5 days in culture. Additionally, more GFS-immunoreactive cells were observed in the connective tissue of the mesentery, specifically in the medial and proximal regions (Figure 7K,L and insets). (Figure 7A–D). Explants cultured for 10 days showed GFS immunoreactive cells in both the coelomic epithelium and the connective tissue of gut rudiment and mesentery, but no GFS immunoreactive nerve fibers were observed (Figure 7M–P). The GFS-immunoreactive cells observed at day 10 of culture displayed the same morphology as those observed at day 1 and 5 of culture. Overall, *in vitro* conditions led to changes in the GFS labeling pattern of cells and fibers in gut explants over time.

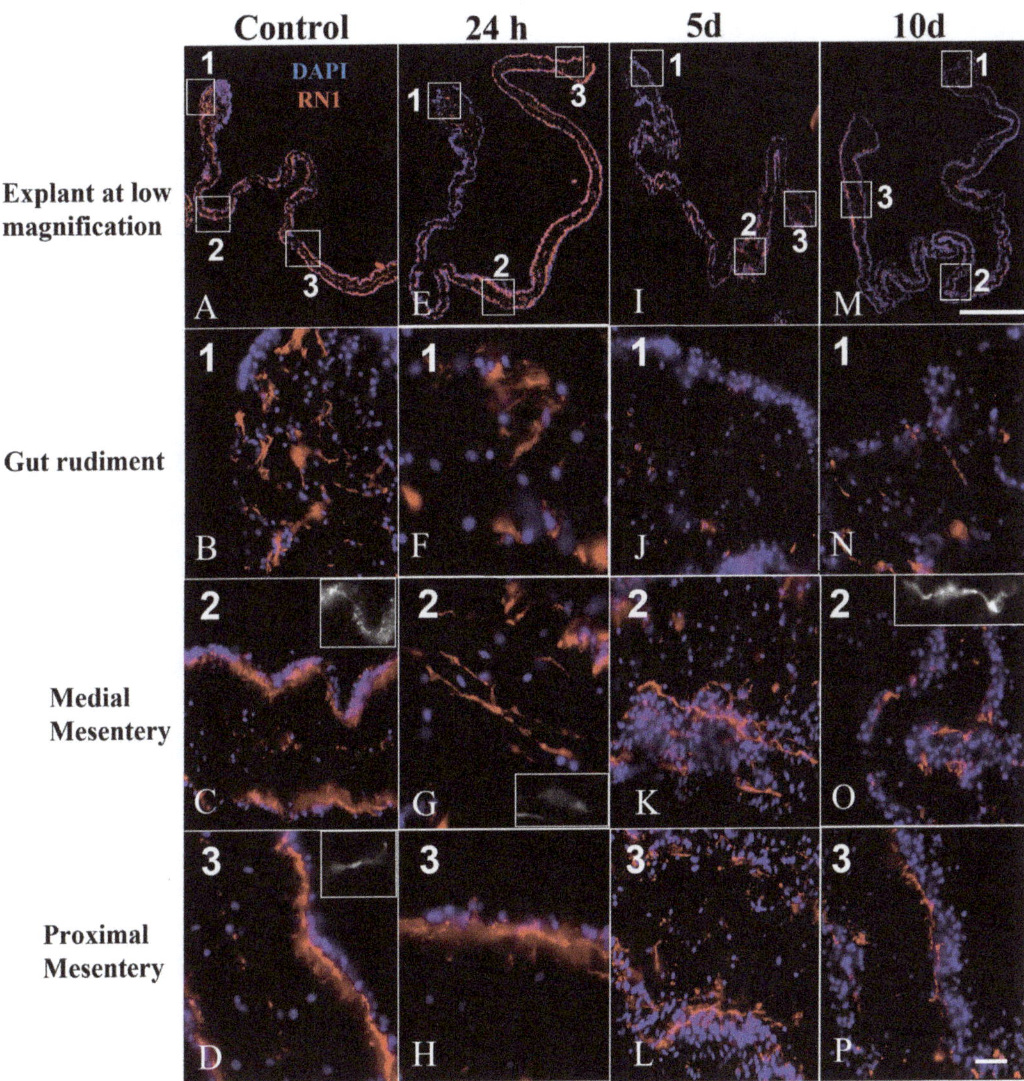

Figure 5. RN1 staining pattern in cultured explants from 5 dpe. RN1 immunoreactivity in control (**A**) and cultured explants for 24 h (**E**), 5 d (**I**), and 10 d (**M**) at low magnification are shown in the upper row. The boxed areas labeled as 1,2, and 3 in these photographs correspond to areas of the gut rudiment (1), medial (2), and proximal mesenteries (3) that are shown at higher magnification from the second to the fourth rows. Asterisks indicate the localization of the gut rudiment in (**A,E,I,M**). RN1 immunoreactive nerve fibers in the gut rudiment of control (**B**) and explants cultured for 24 h (**F**), 5 d (**J**), and 10 d (**N**) are shown at high magnification in the second row. RN1 immunoreactive nerve fibers and neuron-like cell (insets) in the medial mesentery of control (**C**) and explants cultured for 24 h (**G**), 5 d (**K**), and 10 d (**O**) are shown in the third row. RN1 immunoreactive nerve fibers and neuron-like cell (insets) in the proximal mesentery of control (**D**) and explants cultured for 24 h (**H**), 5 d (**L**), and 10 d (**P**) are shown in the bottom row. RN1 immunoreactivity is shown in red and cell nuclei in blue. Bar = (**A,E,I,M**) 500 µm, (**B–D,F–H,J–L,N–P**) 50 µm.

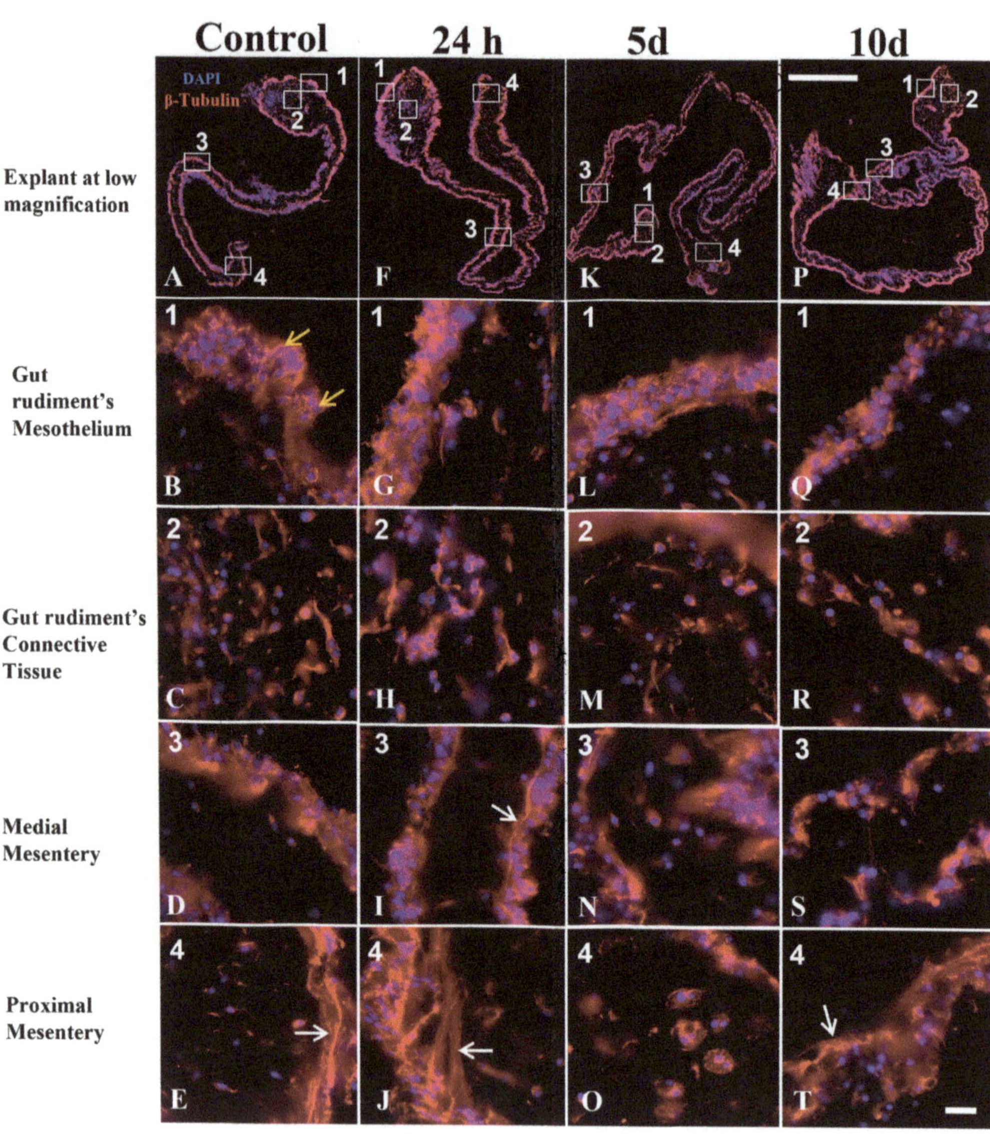

Figure 6. β-tubulin staining pattern in cultured explants from 5 dpe. Control (**A**) and explants cultured for 24 h (**F**), 5 d (**K**), and 10 d (**P**) at low magnification are shown in the upper row. The boxed areas labeled as 1,2, and 3 in these photographs correspond to areas of the gut rudiment (1), medial (2), and proximal mesenteries (3) that are shown at higher magnification from the second to the fourth rows. β-tubulin immunoreactive cells in gut rudiment mesothelium of control ((**B**), yellow arrows), and explants cultured for 24 h (**G**), 5 d (**L**), and 10 d (**Q**) are shown in the second row, while immunoreactive cells in the gut rudiment connective tissue of control (**C**), and explants cultured for 24 h (**H**), 5 d (**M**), and 10 d (**R**) are shown in the third row. β-tubulin immunoreactive cells and fibers in medial mesentery control (**D**) and explants cultured for 24 h (**I**), 5 d (**N**), and 10 d (**S**) are shown in the fourth row, while immunoreactive cells and fibers (white arrows) in the proximal mesentery of control (**E**) and explants cultured for 24 h (**J**), 5 d (**O**), and 10 d (**T**) are shown in the fifth row. β-tubulin immunoreactive cells are shown in red, and cell nuclei labeled with DAPI are shown in blue. Bar = (**A,F,K,P**) 500 μm, (**B–D,G–J,L–O,Q–T**) 50 μm.

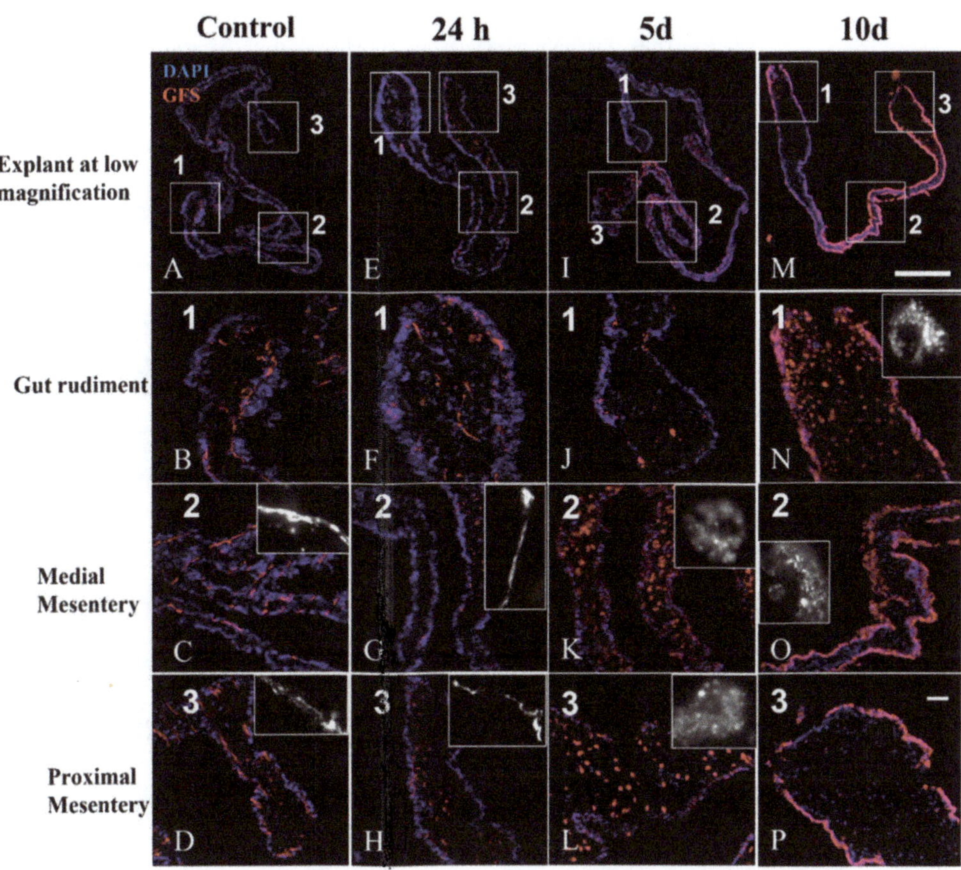

Figure 7. GFS staining pattern in cultured explants from 5 dpe. GFS immunoreactivity in control (**A**) and cultured explants for 24 h (**E**), 5 d (**I**), and 10 d (**M**) at low magnification are shown in the upper row. The boxed areas labeled as 1, 2, and 3 are shown in the second, third, and fourth rows, respectively, at high magnification. GFS- immunoreactive nerve fibers or cells found in the gut rudiment of control (**B**) and explants cultured for 24 h (**F**), 5 d (**J**), and 10 d (**N**) are shown at high magnification in the second row. GFS immunoreactive nerve fibers and cells in the medial mesentery of control (**C**) and explants cultured for 24 h (**G**), 5 d (**K**), and 10 d (**O**) are shown in the third row. GFS immunoreactive nerve fibers and cells in the proximal mesentery of control (**D**) and explants cultured for 24 h (**H**), 5 d (**L**), and 10 d (**P**) are shown in the bottom row. GFS immunoreactivity is shown in red and cell nuclei (DAPI) in blue. Insets show in more detail the GFS-immunoreactive fibers (panels **C,D,G,H**) or cells (panels **K,L,O,P**). Bar = (**A,E,I,M**) 500 µm, (**B–D,F–H,J–L,N–P**) 100 µm.

3.6. Components of the Connective Tissue in Cultured Explants

Fibrillar collagen. The monoclonal antibody Hg-fCOL, which binds specifically to fibrillar collagen, has been used to show how the ECM undergoes dramatic changes during regeneration [22]. The fibrous collagen disappears from the gut rudiment and later from all the mesentery as intestine regeneration advances (at 7 dpe, collagen is not observed in the gut rudiment nor in the mesentery). Later on, after lumen appearance in the newly formed intestine, collagen reappears. Here, we found that, as expected for a regenerating intestine at 5 dpe, the fibrillar collagen had disappeared in certain areas of the gut rudiment but was present all along the mesentery in control explants (Figure 8A–D). A similar distribution of collagen was observed in explants cultured for 24 h (Figure 8E–H), 5 d (Figure 8I–L), or 10 d

(Figure 8M–P), suggesting that the *in vitro* conditions did not change the localization of the collagen. Thus, our findings suggest that the histological characteristics of the extracellular matrix in cultured gut explants remained similar to the 5 dpe *in vivo* counterpart.

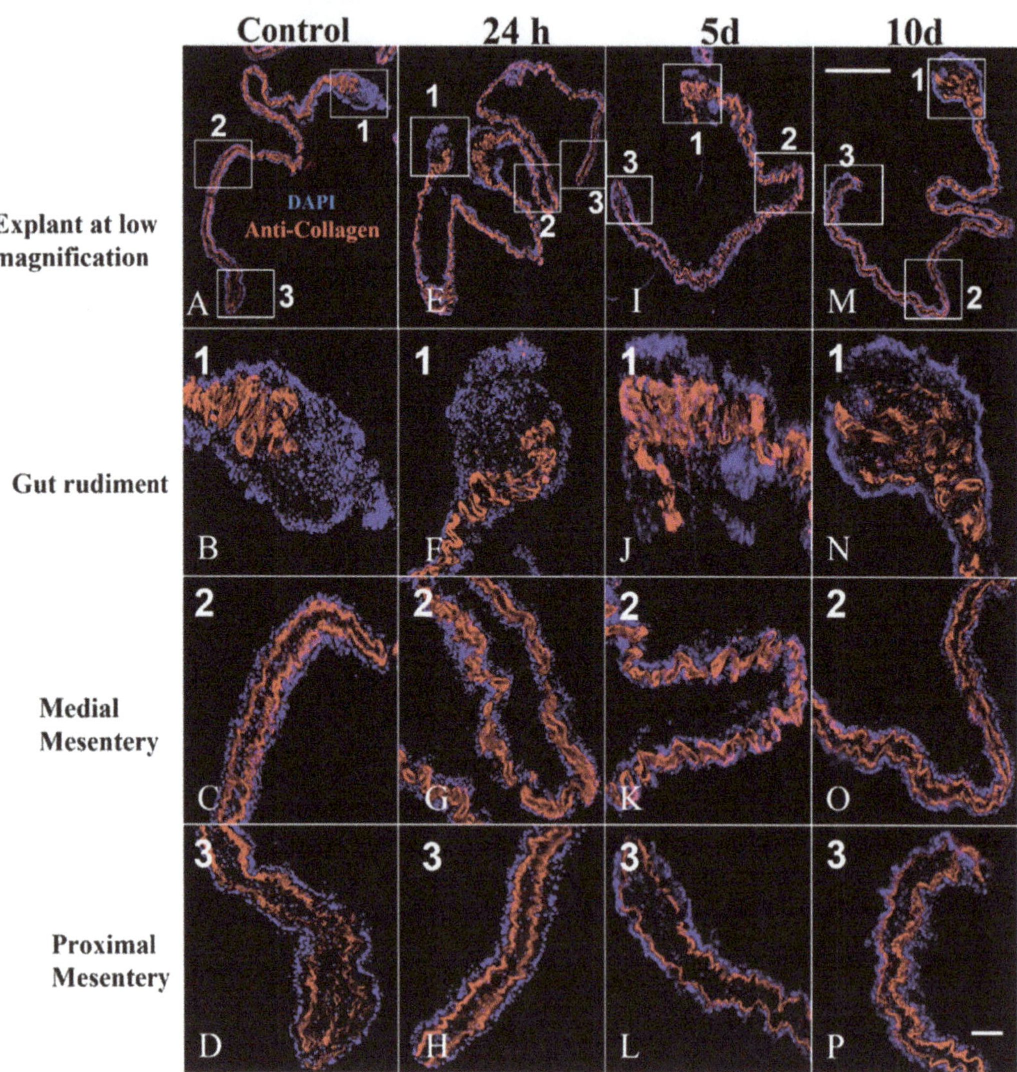

Figure 8. Anti-collagen immunoreactivity in cultured explants dissected at 5 dpe. Anti-collagen immunoreactivity is shown in control explants (**A**) and explants cultured for 24 h (**E**), 5 d (**I**), and 10 d (**M**) at low magnification. The boxed areas labeled as 1, 2, and 3 are shown in the second, third, and fourth rows, respectively, at high magnification. Immunoreactive fibrous collagen is shown at a higher magnification in the gut rudiment of control explants (**B**), and explants cultured for 24 h (**F**), 5 d (**J**), and 10 d (**N**). Micrographs showing collagen immunoreactivity in the medial and proximal regions of the mesentery of control explants are shown in (**C,D**), respectively; explants cultured for 24 h are shown in (**G,H**), respectively; explants cultured for 5 d are shown in (**K,L**) respectively; and in explants cultured for 10 d are shown in (**O,P**), respectively. Collagen immunoreactivity is shown in red and cell nuclei (DAPI) in blue. Bar = (**A,E,I,M**) 500 μm, (**B–D,F–H,J–L,N–P**) 100 μm.

Spherulocytes or morula cells. The monoclonal antibody Sph3 was also developed in our laboratory as a marker for spherulocytes or morula cells; a subpopulation of cells found in the connective tissue of holothurians tissues [21]. Here, we found scattered cells labeled with the Sph3 antibody in control explants (Figure 9A–D), as well as in explants cultured for 24 h (Figure 9E–H), 5 d (Figure 9I–L), and 10 d (Figure 9M–P), following no particular pattern of distribution. The Sph3- immunoreactive cells observed in both control and cultured explants displayed the characteristic morphology of spherulocytes: round or oval with numerous spherule-like structures in their interior (insets in Figure 9C,G,K,O).

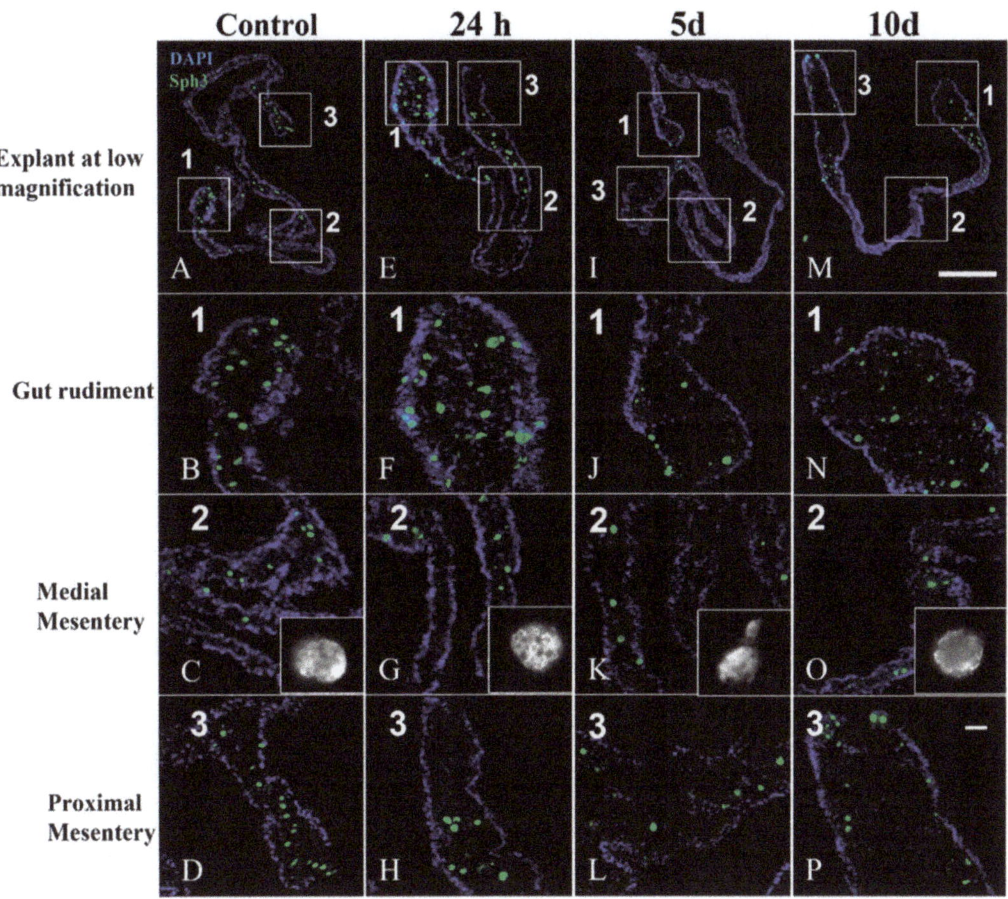

Figure 9. Sph3 staining pattern in cultured explants from 5 dpe. Sph3 immunoreactivity in control (**A**) and cultured explants for 24 h (**E**), 5 d (**I**), and 10 d (**M**) at low magnification are shown in the upper row. The boxed areas labeled as 1, 2, and 3 are shown in the second, third, and fourth rows, respectively, at high magnification. Immunoreactive cells for Sph3 in the gut rudiment of control (**B**) and explants cultured for 24 h (**F**), 5 d (**J**), and 10 d (**N**) are shown at high magnification in the second row. Sph3 immunoreactive cells in the medial mesentery of control (**C**) and explants cultured for 24 h (**G**), 5 d (**K**), and 10 d (**O**) are shown in the third row. Sph3 immunoreactive cells in the proximal mesentery of control (**D**) and explants cultured for 24 h (**H**), 5 d (**L**), and 10 d (**P**) are shown in the lower row. Sph3 immunoreactivity is shown in green and cell nuclei in blue. The insets in (**C,G,K,O**) show the morphology of the spherulocytes. Bar = (**A,E,I,M**) 500 µm, (**B–D,F–H,J–L,N–P**) 100 µm.

3.7. Double Labeling (GFS and Sph3) in Cultured Explants

As both GFS and Sph3 antibodies label cells resembling spherulocytes, we decided to perform a double labeling of cultured tissues, to determine if the spherulocyte-like cells were co-labeled with both antibodies, which could suggest that they correspond to the same cell population. We found that no co-labeling of Sph3 and GFS was observed in control or cultured explants. In control explants only nerve fibers were immunoreactive for GFS (Figure 10A, white arrow), while ovoid cells were immunoreactive for Sph3 (Figure 10A, asterisk). In explants cultured for 24 h, nerve fibers and irregular shaped cells immunoreactive for GFS were observed (Figure 10B, white and yellow arrows, respectively). Additionally, ovoid Sph3-immunoreactive cells were observed (Figure 10B, asterisk). These cells were not co-labeled with GFS and Sph3 antibodies. GFS positive cells, both at the mesothelial and connective tissue, were found in explants cultured for 5 days. Few Sph3 cells were observed in the connective tissue (Figure 10C). After 10 days in culture, a strong GFS immunoreactivity in the mesenterial mesothelium was observed. Additionally, cells immunoreactive for GFS or Sph3 were observed in the connective tissue (Figure 10D). Remarkably, each antibody appears to label a different subset of cells. Overall, the appearance and increase in the number of GFS-immunoreactive cells with culture time correlated with the disappearance of GFS-positive nerve fibers. The GFS-labeled cells were not co-labeled with the Sph3 antibody.

3.8. Explant Outgrowths

Gut explants in organ culture did not adhere firmly to the culture substrates (glass or polystyrene) and remained on them, but not truly attached. Nevertheless, many cells could detach or migrate out from these explants into the culture medium. After 24 h in culture, most of the detached cells, putative coelomocytes according to their spherical or ovoid cell morphology, remained in suspension near the explants (Figure 11A, black arrow). Other cells, spindle-shaped or irregular in morphology, were able to adhere firmly to the substrates and appeared flattened and exhibited cell projections (filopodia) (Figure 11B,C, blue and yellow arrows). After 5 days in culture, spherical cells in suspension were still observed around or near to the explants (Figure 11D, black arrow); however, more cells were flattened and appeared attached to the substrate contacting each other through cell projections (Figure 11E,F, blue and yellow arrows). Remarkably, they did not form a confluent monolayer (Figure 11E). After 10 days in culture spherical cells in suspension were still observed around or near the explants (Figure 11G, black arrow). However, the number of cells attached to the substrate (displaying filopodia and lamellipodia) increased (Figure 11I, yellow arrows). They were able to form net-like structures (Figure 11H).

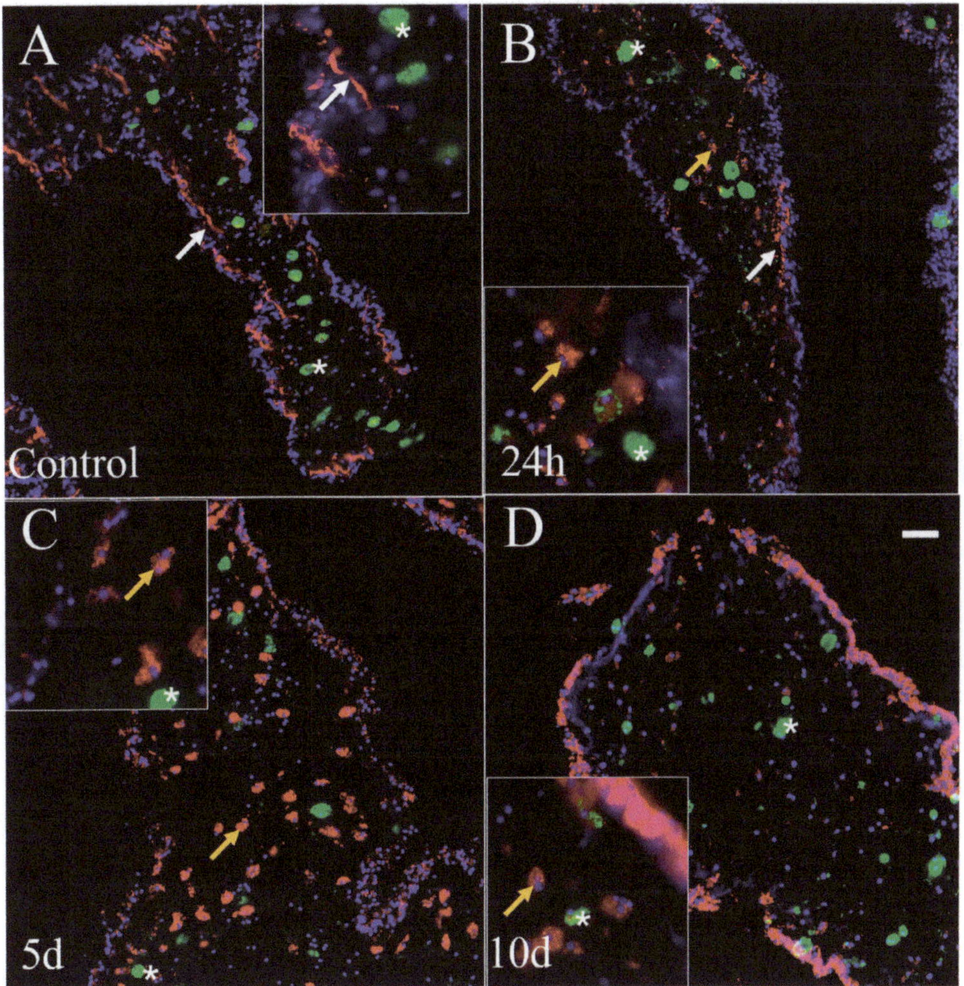

Figure 10. Double labeling (GFS-Sph3) of cultured explants dissected at 5 dpe. The co-labeling of GFS and Sph3 was evaluated in the proximal mesentery of control (**A**) and explants cultured for 24 h (**B**), 5 d (**C**), and 10 d (**D**). Nerve fibers, but not cells, immunoreactive for GFS are observed (white arrows) in controls explants. Sph3 positive cells, which are not co-labeled with GFS, are also observed (asterisks) (**A**). Yellow arrows show GFS positive cells in the connective tissue of explants cultured for 24 h. Asterisks point to Sph3 positive cells (**B**). Many GFS positive cells are observed in the connective tissue of explants cultured for 5 days (yellow arrows), while few Sph3 cells are also present (asterisk) (**C**). Remarkably, the mesenteric mesothelium appears to be immunoreactive for GFS in explants cultured for 10 days. However, GFS (yellow arrows) and Sph3 (asterisks) immunoreactive cells are also observed in explants cultured for 10d (**D**). GFS immunoreactivity is shown in red, Sph3 immunoreactivity is shown in green, and cell nuclei (DAPI) in blue. Bar = (**A–D**) 100 μm.

Phase Contrast　　　　　　　SEM

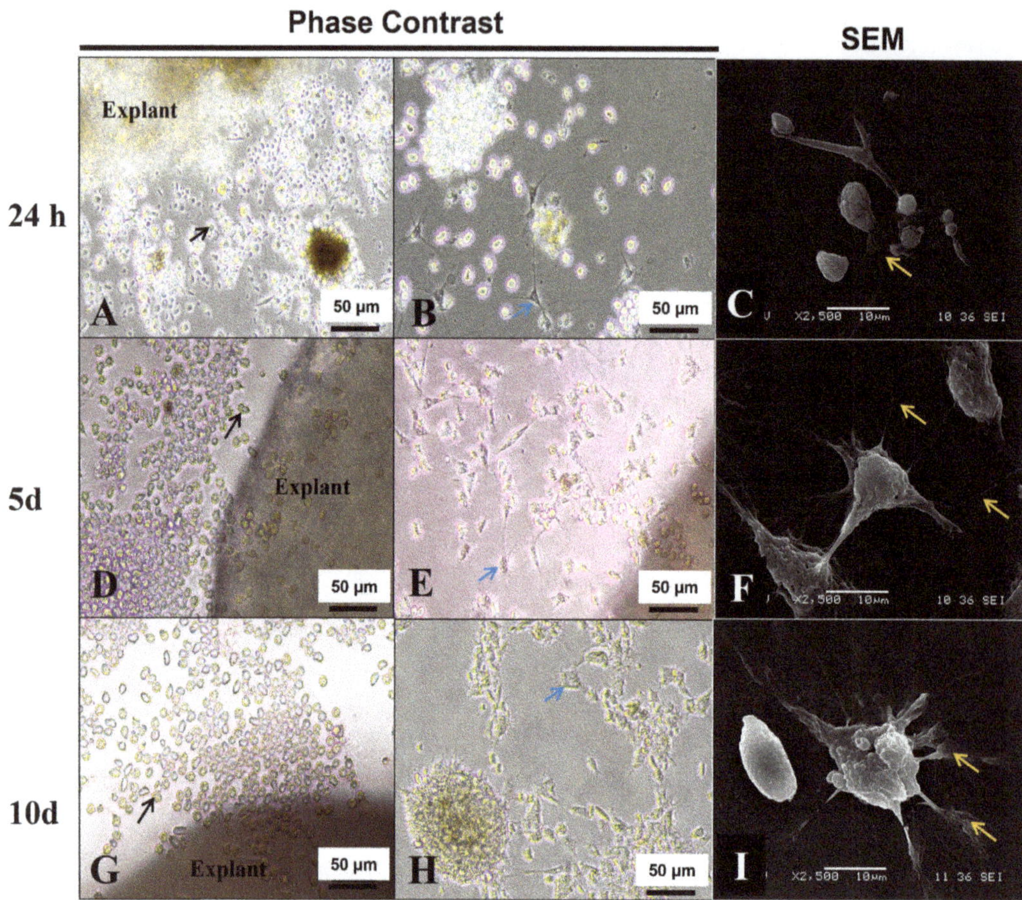

Figure 11. Morphology of cells that migrated out from cultured explants. Representative micrographs of cells that were able to migrate out from explants cultured for 24 h (**A–C**), 5 d (**D–F**), and 10 d (**G–I**) are presented. Micrographs in the first column (**A,D,G**) show the spherical cells located near the explants (black arrows), whereas the second column (**B,E,H**) shows the spindle-shaped or flattened cells (blue arrows) that were able to attach to the culture substrate at the different time points. The third column (**C,F,I**) are SEM micrographs showing the morphology of cells attached to glass cover slips through filopodia and lamellipodia (yellow arrows). Bar = (**A,B,D,E,G,H**) 50 μm, (**C,F,I**) 10 μm.

4. Discussion

Explants in organ culture represent a valuable tool to study the cellular processes underlying developmental or regenerative phenomena in animal models. Explants retain their original histological characteristics; for that reason, they resemble more closely the *in vivo* conditions than do the two-dimension (2-D) culture systems (cell monolayers) [27]. Thus, the response to external stimuli of explant cells is more similar to the response of cells *in vivo*, compared to those of cell monolayers. These differences may be due to the absence of the native extracellular matrix in 2-D culture systems. Components of the extracellular matrix have been recognized as crucial modulators of different physiological processes [28]. Explants in organ culture from both invertebrate and vertebrate organisms have been used to perform several physiological, biochemical, and biomechanical studies [29–32]. For example, in cultured gut explants from mice embryos, natural peristalsis, and stem cell differentiation toward Paneth cells, goblet cells, enterocytes, and others were observed.

Additionally, the effect of modulators of signaling pathways on gut differentiation was studied using gut explants, supporting their value as a research tool [32].

To determine the suitability of cultured explants for studying the effect of external stimuli at the histological level, it is necessary to evaluate the changes that the *in vitro* conditions cause per se on explants. Here, we characterized by histological and immuno-histochemical methods the gut explants dissected from H. glaberrima at 5 dpe and that were maintained in culture up to 10 days. Remarkably, gut explants cultured for 24 h up to 10 days retained their histological characteristics, and each explant's gut rudiment and mesentery could be observed. Key aspects, such as the polarity in the distribution of SLSs and muscle fibers, mesothelium thickness and integrity, and the presence and distribution of fibrous collagen components, are preserved in culture explants as observed in *in vivo* tissues.

However, a few histological changes in the cultured explants compared to fresh explants (controls) and *in vivo* tissues were observed as the length of time in culture increased. Among these were a decrease in nervous system labeling, which might indicate a possible loss over time of nervous system components, the appearance of a putative spherulocyte population, and a decrease in proliferation and loss in the polarity of the proliferating cells. These results agree with those found by other research groups [33,34], where the histological characteristics of retinal explants were preserved up to 14 days, although some changes occurred in a time-dependent manner.

4.1. RN1 Labeling Suggests That a Subpopulation of Nerve Fibers Is Reduced in Cultured Explants as the Time in Culture Increases

Here, we used three different markers previously shown to label nerve fibers and neuron-like cells in the intestinal system of *H. glaberrima*, β-tubulin [17], GFS [20,26], and RN1 [18]. Based on qualitative observations, the overall impression is that the nervous system component is present in the explant in culture but that some components appear to decrease with culture time: β-tubulin labeled components are maintained, while GFS neuronal fibers appear to decrease. Similarly, RN1 labeling suggests that the number of nerve fibers decreases in culture explants. These discrepancies might be explained because each of these three markers recognizes a different subset of nerve fibers and cells in holothurian tissues. Additionally, GFS and anti-β-tubulin are not exclusive to the nervous component; the serum against GFS, as shown previously and in the present manuscript, also labels a spherulocyte-like cell subpopulation [26], while anti-β-tubulin also labels the coelomic epithelial cells [17]. Thus, any quantitative analysis (based on fluorescence intensity) performed with β-tubulin and GFS will be confounded by the labeling of non-neuronal components. Finally, with the use of any of these antibody markers, it is challenging to determine if the decrease in immunofluorescence is due to a decrease in the cell/fiber or a decrease in the expression of the protein. Further efforts must be directed to quantifying by histomorphometry, changes in the presence and number of fibers and neuron-like cells in cultured explants at different culture times, to corroborate the qualitative observations presented here.

4.2. Anti-GFS Labels a Population of Cells That Resemble Spherulocytes in the Connective Tissue and Mesothelial Cells of Cultured Gut Explants

Anti-GFS labels neuron-like cells and nerve fibers in both normal (non-regenerating) and regenerating gut tissues. Interestingly, GFS immunoreactive spherulocyte-like cells were observed after the explants were cultured for 5 days. The GFS immunoreactive cells were observed in both the connective tissue and the mesothelium. The GFS positive cells observed in cultured explants did not appear to be neuron-like cells based solely on their morphology, as they did not exhibit cell projections. In fact, morphologically, they resembled spherulocytes. However, they were not co-labeled with the Sph3 antibody, which is a marker for spherule-containing cells in *H. glaberrima*. The GFS immunoreactive cells observed in cultured explants may correspond to a subpopulation of spherylocytes that are not labeled by Sph3. These spherulocytes might originate from the dedifferentiated mesothelial cells that ingress into the connective tissue. Previous studies have suggested that spherulocytes are derived from coelomic cells [35] and that dedifferentiated mesothelial cells are stem-like cells with a high degree of plasticity [16]. The *in vitro* conditions may induce the differentiation of stem-like cells toward the subpopulation of spherule containing cells, which might participate in the remodeling of the extracellular matrix during the intestine regrowth or might represent a response to the *in vitro* conditions.

Interestingly, the presumptive spherulocytes labeled with the anti-GFS antibody increased with culture time, while those labeled with Sph3 remained relatively constant. These findings suggest that the presumptive spherulocytes labeled with GFS did not originate from the Sph3 subpopulation of cells by a process that involves a change of the vesicular content with a concomitant change of the immunoreactivity. This supports the hypothesis that they are newly formed in cultured explants from dedifferentiated mesothelial cells.

The anti-GFS antibody is a polyclonal antibody produced in rabbits against the synthetic peptide Gly-Phe-Ser-Lys-Leu-Tyr-Phe-NH2 (GFSKLYFamide), which corresponds to the sequence of a neuropeptide isolated from the digestive system of *H. glaberrima* [26]. In previous studies neuron-like cells and nerve fibers, as well as spherule-containing cells immunoreactive to anti-GFS, were described in *H. glaberrima* intestines [20,26]. However, when the serum containing the anti-GFS antibody was pre-absorbed against the synthetic GFSKLYFamide peptide, the immunoreactivity against the neuron-like cells and nerve fibers disappeared, while the immunoreactivity against the spherule-containing cells remained, suggesting that other antibodies different to anti-GFS were present in the rabbit serum [26]. Thus, the labeling of a subpopulation of spherulocyes could be due to non-specific labeling and not necessarily because the cells are expressing GFS. Nonetheless, the fact that a subpopulation of spherulocyte-like cells appears due to the culture conditions, continues to be an interesting phenomenon that must be addressed in future studies.

4.3. The Polarity in the Localization of BrdU-Positive Cell Nuclei Is Lost and the Proliferation Rate Decreases over Time in Cultured Explants

BrdU-positive cells were observed in cultured explants, suggesting that explants in *in vitro* conditions retain their ability to divide as they do in *in vivo* conditions. Interestingly, after 24 h of culture, the BrdU-positive nuclei were mainly localized to the gut rudiment and the distal mesentery, the same sites where the BrdU-positive cells are observed in *in vivo* tissues at 5 dpe [16]. However, in explants cultured for 5 or 10 days this pattern changed dramatically; the BrdU-positive cells were observed scattered in all the regions of the gut rudiment and the mesentery. This finding might be due to changes in the expression of molecules that confer polarity to the explants, such as components of the Frizzled/Planar Cell Polarity and Anterior–Posterior (AP) patterning systems [36] or to the loss of paracrine signaling in the explants in culture over time. Our results do not agree with those of Candia-Carnevali et al. [3], who studied the distribution of BrdU-positive cells in double amputated arm explants from the crinoid *A. meditteranea* after different times in culture. They showed that the BrdU cells were localized mainly in the distal end (where the blastema was formed), the coelomic epithelium and the brachial nerve of the double amputated arm explants. Remarkably, this pattern persisted in explants cultured from 24 h for up to one week.

The proliferation rate in the gut rudiment of explants from 5 dpe cultured for 24 h was similar to that observed *in vivo* (in the range of 15 to 16%). Remarkably, the proliferation rate was lower after 10 d in culture displaying a trend toward a reduction over time. These results agree with those of other groups [37–39], who observed that the proliferation rate decreased over time in mouse kidney, human colon, and heart zebrafish explants, respectively. The causes of the decline in cell proliferation could be missing ingredients in the culture medium, loss of paracrine signaling, or suboptimal environmental conditions. Despite this limitation, the gut explant culture described here is suitable for performing short-term studies on cell proliferation.

Interestingly, while the polarity in the localization of proliferating cells in explants is lost as the time in culture advances, the polarity in the distribution of SLSs and muscle fibers is maintained in explants. These findings support our lab's hypothesis that cell dedifferentiation and proliferation are uncoupled, and controlled by different signaling pathways at the early stages of intestine regeneration in *H. glaberrima* [8]. Previous studies by our group found that cell proliferation appears to be controlled by the canonical Wnt pathway [8,11], while muscle dedifferentiation appears to be modulated by GSK-3 signaling in a Wnt-independent manner [8]. Thus, culture conditions might affect the paracrine signaling of the canonical Wnt pathway but not the pathway(s) modulating the muscle dedifferentiation. Further studies must address this topic.

4.4. The Cellular Processes Underlying Intestine Regeneration Are Halted in Cultured Gut Explants Compared to the In Vivo Counterparts

Our findings suggest that gut rudiment size and most of the histological characteristics of cultured gut explants remain similar to those observed in intestines from the same regenerating stage from which they were dissected initially (e.g., 5 dpe), without significant changes over time *in vitro*. Remarkably, the cellular events that generally occur in guts as regeneration proceeds *in vivo* appear to be halted, or at least to occur slower, in the cultured explants. The main feature being that no growth of the rudiment is observed, whereas *in vivo*, the rudiment keeps increasing in size during the second regeneration week [16]. Another example is the lack of muscle fiber reappearance and the lumen formation in the gut rudiment of explants cultured for 10 days. The *in vivo* counterpart of these explants will be gut tissues of 15 dpe (5 dpe plus 10 days of culture), which already have defined muscle layers as constituents of the mesothelium and display a luminal space. In this respect, our findings do not agree with Candia-Carnevali [3] and Dupont and Thorndyke [4], who observed that the regeneration-associated events described *in vivo* proceed in crinoid and ophiuroid arm explants, although at a slower rate. These differences might be because crinoid or ophiuroid arms naturally exist as external structures (in a similar environment to culture conditions). In contrast, the intestine and other viscera regenerate within the coelomic cavity, a more controlled space, where the presence of paracrine signaling molecules could be more crucial for the regeneration to proceed. Thus, missing growth factors could have a more significant effect on the visceral than in appendicular regeneration in echinoderms. An additional component that could contribute to the halting of gut regeneration in the cultured explants is the lack of cells that migrate from other organs to the gut rudiment through the mesentery *in vivo*, where they contribute to the formation of structures such as the lumen. Optimization of the current culture conditions might emulate more accurately the intestine regeneration process in explants, adding value to our *in vitro* model.

Overall, the gut explant culture method described here is suitable for short-term studies on critical aspects of the intestine regeneration process. The presence of mesothelial, neural, and extracellular matrix components in cultured explants similar to those observed in non-cultured explants and *in vivo* tissues [16–22] is a remarkable finding. Explants retain their morphology and histoarchitecture for up to 10 days, with minor changes. Additionally, explants maintain the original polarity in the distribution and localization of muscle bundles, SLSs, and proliferating cells for up to 5 days. Apart from the applicability in histological studies, explants can also be used to perform molecular studies (such as gene expression studies by qRT-PCR) in response to external stimuli [40]. The utility of gut explants has already been demonstrated in studies performed by our group to test the effect of pharmacological agents on muscle dedifferentiation [8], to set up the RNA interference technique to decipher the role of the canonical Wnt signaling pathway in the intestine regeneration in *H. glaberrima* [11,13], and to determine the toxic effects of antibiotics on holothurian tissues [41], demonstrating their value as a tool to complement the *in vivo* studies in echinoderms.

Author Contributions: Conceptualization, S.A.B. and J.E.G.-A.; methodology, S.A.B. and J.E.G.-A.; investigation, S.A.B.; validation, S.A.B. and J.E.G.-A.; formal analysis S.A.B.; resources, S.A.B. and J.E.G.-A.; data curation, S.A.B. and J.E.G.-A.; writing—original draft preparation, S.A.B.; writing— review and editing, S.A.B. and J.E.G.-A.; visualization, S.A.B.; supervision, J.E.G.-A.; project adminis- tration, J.E.G.-A.; funding acquisition, J.E.G.-A. All authors have read and agreed to the published version of the manuscript.

Funding: This project was funded by NIH (Grant R15GM124595 and R21AG057974), NSF (Grant IOS-1252679) and the University of Puerto Rico. S.A.B. was partially funded by the University of Puerto Rico through the Merit and Doctoral Dissertation Fellowships.

Institutional Review Board Statement: This research deals only with invertebrate animals, thus the University of Puerto Rico IACUC waives ethical approval of research performed on invertebrates.

Informed Consent Statement: Not applicable.

Data Availability Statement: The data presented in this study it is contained within this article.

Acknowledgments: Special thanks to Rey Rosa Morales and Griselle Valentín-Tirado for their tech- nical support. We gratefully acknowledge the Materials Characterization Center (MCC) of the University of Puerto Rico for the use of the Scanning Electron Microscopy Facility.

Conflicts of Interest: The authors declare no conflict of interest.

Appendix A

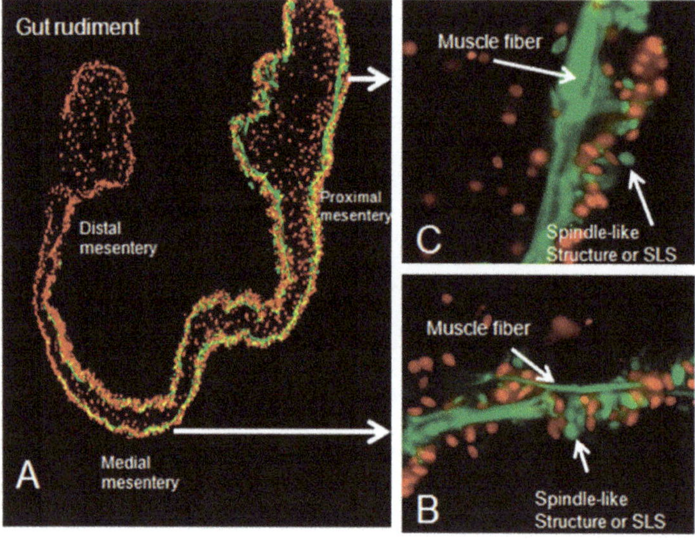

Figure A1. SLSs and muscle gradient in the regenerating mesentery. (**A**) Tissue section of a regener- ating intestine at 5 dpe. The SLSs and muscle fibers have disappeared from the gut rudiment and the distal mesentery, while they are still present in the medial and the proximal mesenteries at this regenerative stage. SLSs and muscle fibers are shown in medial (**B**) and proximal (**C**) mesentery tissue sections. SLSs and muscle fibers were labeled with Phalloidin-TRITC and are displayed in green, and cell nuclei were labeled with DAPI and are displayed in red.

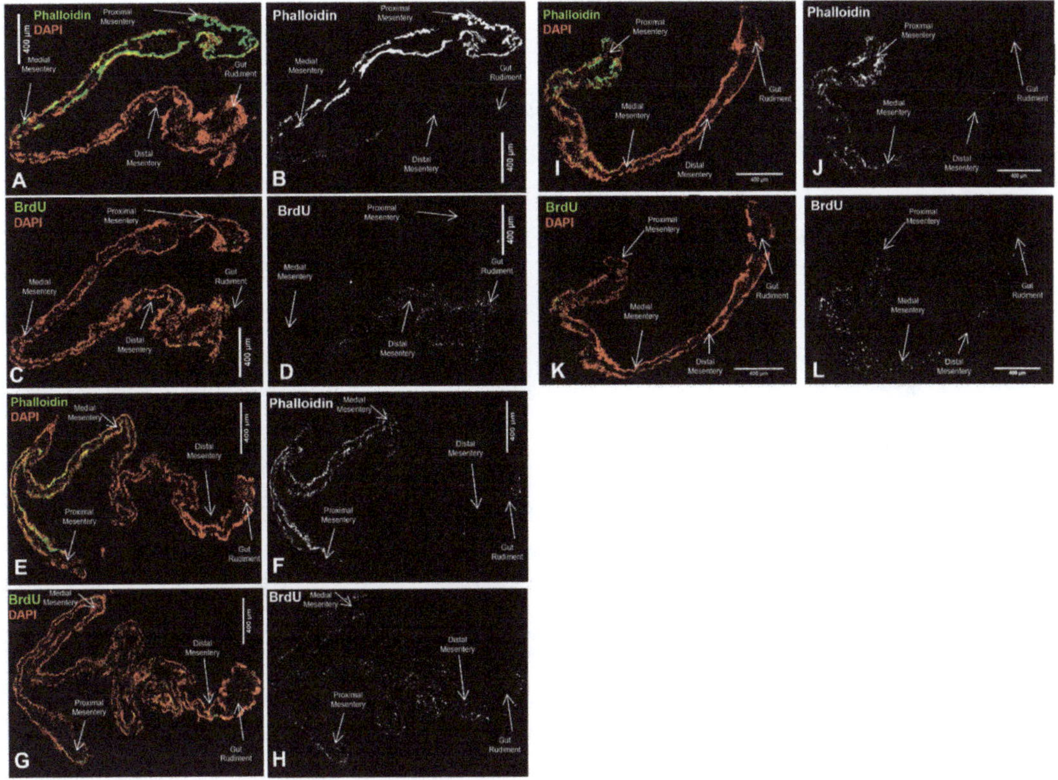

Figure A2. Phalloidin and anti-BrdU staining pattern in cultured explants from 5 dpe. Tissue sections of explants cultured for 24 h (**A–D**), 5 d (**E–H**), and 10 d (**I–L**) were labeled with phalloidin (**A,B**), (**E,F**), and (**I,J**), respectively, or with anti-BrdU (**C,D**), (**G,H**), and (**K,L**), respectively. In each explant the gut rudiment, the distal, medial, and proximal mesentery are signaled. Muscle and SLSs stained with Phalloidin-TRITC and cell nuclei immunoreactive for anti-BrdU are shown in green, while total cell nuclei are shown in red. The panels (**B,D,F,H,J,L**) are shown in white and black. Scale bar = 400 μm.

References

1. Freshney, R.I. (Ed.) Primary Culture. In *Culture of Animal Cells: A Manual of Basic Technique and Specialized Applications*; John Wiley & Sons: Hoboken, NJ, USA, 2010; pp. 163–186. ISBN 978-0-470-52812-9.
2. Yissachar, N.; Zhou, Y.; Ung, L.; Lai, N.Y.; Mohan, J.F.; Ehrlicher, A.; Weitz, D.A.; Kasper, D.L.; Chiu, I.M.; Mathis, D.; et al. An Intestinal Organ Culture System Uncovers a Role for the Nervous System in Microbe-Immune Crosstalk. *Cell* **2017**, *168*, 1135–1148.e12. [CrossRef]
3. Carnevali, C.; Bonasoro, F.; Patruno, M.; Thorndyke, M.C. Cellular and molecular mechanisms of arm regeneration in crinoid echinoderms: The potential of arm explants. *Dev. Genes Evol.* **1998**, *208*, 421–430. [CrossRef]
4. Dupont, S.; Thorndyke, M.C. Growth or differentiation? Adaptive regeneration in the brittlestar Amphiura filiformis. *J. Exp. Biol.* **2006**, *209*, 3873–3881. [CrossRef]
5. Burns, G.; Ortega-Martinez, O.; Dupont, S.; Thorndyke, M.C.; Peck, L.S.; Clark, M.S. Intrinsic gene expression during regeneration in arm explants of Amphiura filiformis. *J. Exp. Mar. Bio. Ecol.* **2012**, *413*, 106–112. [CrossRef]
6. Di Benedetto, C.; Parma, L.; Barbaglio, A.; Sugni, M.; Bonasoro, F.; Candia Carnevali, M.D. Echinoderm regeneration: An *in vitro* approach using the crinoid Antedon mediterranea. *Cell Tissue Res.* **2014**, *358*, 189–201. [CrossRef]
7. Moss, C.; Beesley, P.W.; Thorndyke, M.C.; Bollner, T. Preliminary observations on ascidian and echinoderm neurons and neural explants *in vitro*. *Tissue Cell* **1998**, *30*, 517–524. [CrossRef]
8. Bello, S.A.; Torres-Gutiérrez, V.; Rodríguez-Flores, E.J.; Toledo-Román, E.J.; Rodríguez, N.; Díaz-Díaz, L.M.; Vázquez-Figueroa, L.D.; Cuesta, J.M.; Grillo-Alvarado, V.; Amador, A.; et al. Insights into intestinal regeneration signaling mechanisms. *Dev. Biol.* **2020**, *458*, 12–31. [CrossRef]

9. García-Arrarás, J.E.; Lázaro-Peña, M.I.; Díaz-Balzac, C.A. Holothurians as a Model System to Study Regeneration. In *Results and Problems in Cell Differentiation*; Springer: Cham, Switzerland, 2018; Volume 65, pp. 255–283.

10. García-Arrarás, J.E.; Bello, S.A.; Malavez, S. The mesentery as the epicenter for intestinal regeneration. In *Seminars in Cell & Developmental Biology*; Academic Press: Cambridge, MA, USA, 2018.

11. Alicea-Delgado, M.; García-Arrarás, J.E. Wnt/β-catenin signaling pathway regulates cell proliferation but not muscle dedifferentiation nor apoptosis during sea cucumber intestinal regeneration. *Dev. Biol.* **2021**, *480*, 105–113. [CrossRef]

12. Quesada-Díaz, E.; Figueroa-Delgado, P.; García-Rosario, R.; Sirfa, A.; García-Arrarás, J.E. Dedifferentiation of radial glia-like cells is observed in *in vitro* explants of holothurian radial nerve cord. *J. Neurosci. Methods* **2021**, *364*, 109358. [CrossRef]

13. Alicea-Delgado, M.; Bello-Melo, S.A.; García-Arrarás, J.E. RNA Interference on Regenerating Holothurian Gut Tissues. In *Methods in Molecular Biology*; Humana: New York, NY, USA, 2021; pp. 241–252. ISBN 978-1-0716-0973-6.

14. Bello, S.A.; Abreu-Irizarry, R.J.; García-Arrarás, J.E. Primary cell cultures of regenerating holothurian tissues. *Methods Mol. Biol.* **2015**, *1189*, 283–297.

15. Schacher, S.; Proshansky, E. Neurite regeneration by Aplysia neurons in dissociated cell culture: Modulation by Aplysia hemolymph and the presence of the initial axonal segment. *J. Neurosci.* **1983**, *3*, 2403–2413. [CrossRef]

16. García-Arrarás, J.E.; Valentín-Tirado, G.; Flores, J.E.; Rosa, R.J.; Rivera-Cruz, A.; San Miguel-Ruiz, J.E.; Tossas, K. Cell dedifferentiation and epithelial to mesenchymal transitions during intestinal regeneration in H. glaberrima. *BMC Dev. Biol.* **2011**, *11*, 61. [CrossRef]

17. Tossas, K.; Qi-Huang, S.; Cuyar, E.; García-Arrarás, J.E. Temporal and spatial analysis of enteric nervous system regeneration in the sea cucumber *Holothuria glaberrima*. *Regeneration* **2014**, *1*, 10–26. [CrossRef]

18. Díaz-Balzac, C.A.; Santacana-Laffitte, G.; San Miguel-Ruíz, J.E.; Tossas, K.; Valentín-Tirado, G.; Rives-Sánchez, M.; Mesleh, A.; Torres, I.I.; García-Arrarás, J.E. Identification of nerve plexi in connective tissues of the sea cucumber *Holothuria glaberrima* by using a novel nerve-specific antibody. *Biol. Bull.* **2007**, *213*, 28–42. [CrossRef]

19. Rosado-Olivieri, E.A.; Ramos-Ortiz, G.A.; Hernández-Pasos, J.; Díaz-Balzac, C.A.; Vázquez-Rosa, E.; Valentín-Tirado, G.; Vega, I.E.; García-Arrarás, J.E. A START-domain-containing protein is a novel marker of nervous system components of the sea cucumber *Holothuria glaberrima*. *Comp. Biochem. Physiol. Part-B Biochem. Mol. Biol.* **2017**, *214*, 57–65. [CrossRef]

20. Díaz-Miranda, L.; Blanco, R.E.; García-Arrarás, J.E. Localization of the heptapeptide GFSKLYFamide in the sea cucumber holothuria glaberrima (echinodermata): A light and electron microscopic study. *J. Comp. Neurol.* **1995**, *352*, 626–640. [CrossRef]

21. García-Arrarás, J.E.; Schenk, C.; Rodrígues-Ramírez, R.; Torres, I.I.; Valentín, G.; Candelaria, A.G. Spherulocytes in the echinoderm *Holothuria glaberrima* and their involvement in intestinal regeneration. *Dev. Dyn.* **2006**, *235*, 3259–3267. [CrossRef]

22. Quiñones, J.L.; Rosa, R.; Ruiz, D.L.; García-Arrarás, J.E. Extracellular Matrix Remodeling and Metalloproteinase Involvement during Intestine Regeneration in the Sea Cucumber *Holothuria glaberrima*. *Dev. Biol.* **2002**, *250*, 181–197. [CrossRef]

23. Bello, S.A.; De Jesús-Maldonado, I.; Rosim-Fachini, E.; Sundaram, P.A.; Diffoot-Carlo, N. In vitro evaluation of human osteoblast adhesion to a thermally oxidized γ-TiAl intermetallic alloy of composition Ti-48Al-2Cr-2Nb (at.%). *J. Mater. Sci. Mater. Med.* **2010**, *21*, 1739–1750.

24. Candelaria, A.G.; Murray, G.; File, S.K.; García-Arrarás, J.E. Contribution of mesenterial muscle dedifferentiation to intestine regeneration in the sea cucumber *Holothuria glaberrima*. *Cell Tissue Res.* **2006**, *325*, 55–65. [CrossRef]

25. Dolmatov, I.Y.; Ginanova, T.T. Muscle Regeneration in Holothurians. *Microsc. Res. Tech.* **2001**, *55*, 452–463. [CrossRef]

26. Diaz-Miranda, L.; Price, D.A.; Greenberg, M.J.; Lee, T.D.; Doble, K.E.; Garcia-Arraras, J.E. Characterization of Two Novel Neuropeptides from the Sea Cucumber *Holothuria glaberrima*. *Biol. Bull.* **1992**, *182*, 241–247. [CrossRef]

27. Al-Lamki, R.S.; Bradley, J.R.; Pober, J.S. Human organ culture: Updating the approach to bridge the gap from *in vitro* to *in vivo* in inflammation, cancer, and stem cell biology. *Front. Med.* **2017**, *4*, 148. [CrossRef]

28. Frantz, C.; Stewart, K.M.; Weaver, V.M. The extracellular matrix at a glance. *J. Cell Sci.* **2010**, *123*, 4195–4200. [CrossRef]

29. Koch, M.; Hassan, B.A. The Making and Un-Making of Neuronal Circuits in Drosophila. Out with the Brain: Drosophila Whole-Brain Explant Culture. *Neuromethods* **2012**, *69*, 261–268.

30. Lorenzen, U.S.; Hansen, G.H.; Danielsen, E.M. Organ culture as a model system for studies on Enterotoxin interactions with the intestinal epithelium. *Methods Mol. Biol.* **2016**, *1396*, 159–166.

31. Taylor, V.; Hicks, J.; Ferguson, C.; Willey, J.; Danelson, K. Effects of tissue culture on the biomechanical properties of porcine meniscus explants. *Clin. Biomech.* **2019**, *69*, 120–126. [CrossRef]

32. Sun, X.; Fu, X.; Du, M.; Zhu, M.J. Ex vivo gut culture for studying differentiation and migration of small intestinal epithelial cells. *Open Biol.* **2018**, *4*, 170256. [CrossRef]

33. Johnson, T.V.; Martin, K.R. Development and characterization of an adult retinal explant organotypic tissue culture system as an *in vitro* intraocular stem cell transplantation model. *Investig. Ophthalmol. Vis. Sci.* **2008**, *49*, 3503–3512. [CrossRef]

34. Müller, B.; Wagner, F.; Lorenz, B.; Stieger, K. Organotypic cultures of adult mouse retina: Morphologic changes and gene expression. *Investig. Ophthalmol. Vis. Sci.* **2017**, *58*, 1930–1940. [CrossRef]

35. Byrne, M. The ultrastructure of the morula cells of Eupentacta quinquesemita (Echinodermata: Holothuroidea) and their role in the maintenance of the extracellular matrix. *J. Morphol.* **1986**, *188*, 179–189. [CrossRef]

36. Zallen, J.A. Planar Polarity and Tissue Morphogenesis. *Cell* **2007**, *129*, 1051–1063. [CrossRef]

37. Robertson, H.; Wheeler, J.; Morley, A.R. Culture and *in vitro* bromodeoxyuridine-labelling of mouse kidney explants. *J. Tissue Cult. Methods* **1994**, *16*, 49–55. [CrossRef]

38. Usugane, M.; Fujita, M.; Lipkin, M.; Palmer, R.; Friedman, E.; Augenlicht, L. Cell proliferation in explant cultures of human colon. *Digestion* **1982**, *24*, 225–233. [CrossRef]
39. Cao, J.; Poss, K.D. Explant culture of adult zebrafish hearts for epicardial regeneration studies. *Nat. Protoc.* **2016**, *11*, 872–881. [CrossRef]
40. Bello-Melo, S.A. Determination of the Expression Levels of *axin*, *β-catenin* and *myc* Genes by Quantitative Real Time PCR in Explants Treated with Modulators of the Wnt Signaling Pathway. In Setting-Up an In Vitro Model to Study the Signaling Mechanisms Associated with Intestine Regeneration in the Sea Cucumber *Holothuria glaberrima*. Ph.D. Thesis, Rio Piedras Campus, University of Puerto Rico, San Juan, PR, USA, 2019; pp. 216–220.
41. Díaz-Díaz, L.M.; Rosario-Meléndez, N.; Rodríguez-Villafañe, A.; Figueroa-Vega, Y.Y.; Pérez-Villafañe, O.A.; Colón-Cruz, A.M.; Rodríguez-Sánchez, P.I.; Cuevas-Cruz, J.M.; Malavez-Cajigas, S.J.; Maldonado-Chaar, S.M.; et al. Antibiotics Modulate Intestinal Regeneration. *Biology* **2021**, *10*, 236. [CrossRef]

Journal of
Marine Science and Engineering

Biological and Proteomic Characterization of the Anti-Cancer Potency of Aqueous Extracts from Cell-Free Coelomic Fluid of *Arbacia lixula* Sea Urchin in an In Vitro Model of Human Hepatocellular Carcinoma

Claudio Luparello [1,*], Rossella Branni [1], Giulia Abruscato [1], Valentina Lazzara [1], Simon Sugár [2], Vincenzo Arizza [1], Manuela Mauro [1], Vita Di Stefano [1] and Mirella Vazzana [1]

1   Dipartimento di Scienze e Tecnologie Biologiche, Chimiche e Farmaceutiche (STEBICEF), Università Di Palermo, Viale Delle Scienze, 90128 Palermo, Italy
2   MS Proteomics Research Group, Research Centre for Natural Sciences, Eötvös Loránd Research Network, 1117 Budapest, Hungary
*   Correspondence: claudio.luparello@unipa.it; Tel.: +39-91-238-97405

Citation: Luparello, C.; Branni, R.; Abruscato, G.; Lazzara, V.; Sugár, S.; Arizza, V.; Mauro, M.; Di Stefano, V.; Vazzana, M. Biological and Proteomic Characterization of the Anti-Cancer Potency of Aqueous Extracts from Cell-Free Coelomic Fluid of *Arbacia lixula* Sea Urchin in an In Vitro Model of Human Hepatocellular Carcinoma. *J. Mar. Sci. Eng.* **2022**, *10*, 1292. https://doi.org/10.3390/jmse10091292

Academic Editor: Alexandre Lobo-da-Cunha

Received: 28 July 2022
Accepted: 9 September 2022
Published: 13 September 2022

Publisher's Note: MDPI stays neutral with regard to jurisdictional claims in published maps and institutional affiliations.

Abstract: Echinoderms are an acknowledged source of bioactive compounds exerting various beneficial effects on human health. Here, we examined the potential in vitro anti-hepatocarcinoma effects of aqueous extracts of the cell-free coelomic fluid obtained from the sea urchin *Arbacia lixula* using the HepG2 cell line as a model system. This was accomplished by employing a combination of colorimetric, microscopic and flow cytometric assays to determine cell viability, cell cycle distribution, the possible onset of apoptosis, the accumulation rate of acidic vesicular organelles, mitochondrial polarization, cell redox state and cell locomotory ability. The obtained data show that exposed HepG2 cells underwent inhibition of cell viability with impairment of cell cycle progress coupled to the onset of apoptotic death, the induction of mitochondrial depolarization, the inhibition of reactive oxygen species production and acidic vesicular organelle accumulation, and the block of cell motile attitude. We also performed a proteomic analysis of the coelomic fluid extract identifying a number of proteins that are plausibly responsible for anti-cancer effects. Therefore, the anti-hepatocarcinoma potentiality of *A. lixula*'s preparation can be taken into consideration for further studies aimed at the characterization of the molecular mechanism of cytotoxicity and the development of novel prevention and/or treatment agents.

Keywords: coelomic fluid; sea urchin; echinoderm; HepG2 cells; apoptosis; cell cycle; acidic vesicular organelles; mitochondrial transmembrane potential; reactive oxygen species; wound healing assay

1. Introduction

The marine environment, which occurs in about 70% of the globe's surface and 90% of the biosphere, is the largest habitat of the Earth. It harbors a significant fraction of the world's biodiversity with many still unknown species and represents a little exploited treasure chest rich in bioactive natural products. In fact, animal adaptation processes to the different and often extreme aquatic environments have led to the establishment of unique bio-synthetic pathways [1]. Within this context, the development of interindividual signalization systems and defensive strategies, e.g., against predators, infectious agents and UV radiations, prompted sessile aquatic species to set up networks of chemical communication through a variety of exclusive marine species-specific metabolites, which can be available for extraction and characterization as a massive library of natural active compounds [2,3].

One of the acknowledged potential applications shared among several marine natural products is related to cancer prevention and treatment. Studies performed on compounds extracted from sponges, mollusks, soft corals and tunicates demonstrated their relevant cytotoxic and anti-tumoral properties against different human cancer cell lines in vitro, also acting upon disparate cellular functions such as locomotory ability, gene expression reprogramming and autophagic flux [4–8]. Notably, based on their pharmacological properties and added health benefits, marine bioactives are increasingly considered promising additives for the development of functional foods and active food-packaging material [9–11].

Among the marine invertebrates, the phylum Echinodermata (Klein, 1778) is widely distributed worldwide, consisting of about 7000 extant species grouped into five well-defined taxonomic classes. The non-edible sea urchin *Arbacia lixula* (Linnaeus, 1758; Echinoidea, Arbacioida: Arbaciidae), which colonized the Mediterranean shallow rocks during the Pleistocene [12], is a warm-temperate water organism that has been used as a model for chemical embryotoxicity studies and to determine the effects of high-frequency noises on the immune response [13–15]. The data dealing with the biomedical applications of *A. lixula*'s components are limited to date. In this regard, *A. lixula*'s eggs have been proven as a valuable source of the free radical scavenger astaxanthin, whereas its coelomic fluid and the cells contained therein were found to possess a strong anti-oxidant activity [16–18]. In addition, potentially anti-microbial peptides were also identified in the coelomic fluid of this organism [19].

We previously demonstrated that the cell-free coelomic fluid extract (CFE) obtained from this species is able to exert cytotoxic effects against triple-negative breast cancer cells in culture [20], thus prompting the extension of such in vitro trials also to other tumor cell model systems. The physiological benefits of potential additives for functional foods or food-packaging purposes include the maintenance of the health of the organs of the digestive apparatus. Thus, we chose to expand the knowledge on the potential anti-tumoral effects of *A. lixula*'s CFE by focusing our interest on an in vitro model system representative of tumor-affected liver parenchyma. For this purpose, we studied the biological effects exerted on the HepG2 cell line derived from liver biopsies of a 15-year-old Caucasian male suffering from a differentiated hepatocellular carcinoma [21]. It is widely acknowledged that the proper characterization of the biological activity of new potential anti-cancer products is valuable to establish their efficacy profiles and discover new applications. Moreover, information on their mechanisms of action may reveal new molecular targets of beneficial effect, ultimately serving as a link to clinical activity by designing appropriate interventions and therapies for specific tumor cytotypes [22]. Therefore, the aim of our investigation was to examine the CFE-mediated effects on cell viability and to obtain more insight into the intracellular mechanisms of cytotoxic induction by employing a combination of colorimetric, microscopic and flow cytometric assays to determine cell cycle distribution, the possible onset of apoptosis, the accumulation rate of acidic vesicular organelles (AVOs), mitochondrial polarization and cell redox state and cell locomotory ability. The obtained data show that the CFE treatment of HepG2 cells caused the inhibition of cell viability with impairment of cell cycle progress coupled to the onset of apoptotic death, the induction of mitochondrial depolarization, the inhibition of reactive oxygen species (ROS) production and AVO accumulation, and the impairment of cell motile attitude. As a complement to the phenotypic characterization, the data obtained from the proteomic analysis of the CFE identified a number of proteins plausibly responsible for the observed impairment of biological activity in HepG2 cells.

2. Materials and Methods

2.1. Catching and Maintaining the Animals

A total of 60 individuals of sea urchins of the species *A. lixula* (Figure 1) were collected from the rocky coves of the Gulf of Palermo (Sicily, Italy) at a depth of 5–10 m. Before starting the experiments, the animals were acclimatized in aquaria for two weeks in recirculated and filtered artificial seawater (9 mM KCl; 0.425 M NaCl; 0.0255 M $MgSO_4 \cdot 7H_2O$; 9.3 mM $CaCl_2 \cdot 2H_2O$; 0.023 M $MgCl_2 \cdot 6H_2O$; 2 mM $NaHCO_3$, pH 8.0) at $15 \pm 2\ ^\circ C$ with constant flow-through of dissolved O_2 and fed with commercially available invertebrate food (Algamac 3000, Aquafauna Bio-Marine Inc., Hawthorne, CA, USA).

Figure 1. An *A. lixula* sea urchin specimen.

2.2. Bleeding Procedure and Preparation of the CFE

As already reported by Luparello et al. [23], the CF was collected by cutting the peristomial membrane of the sea urchins and harvesting the bleeding fluid in the presence of cold 60ISO-EDTA anticoagulant (20 mM Tris, 0.5 mM NaCl, 70 mM EDTA; pH 7.5). The CF was immediately centrifuged at $1000\times g$ for 10 min at 4 °C to obtain a cell-free extract, which was stored at -80 °C and subsequently lyophilized in an Alpha 2–4 LD plus freeze-dryer (Martin Christ, Osterode am Harz, D). The lyophilized sample was resuspended in the minimum volume of sterile distilled water and kept at -20 °C until used for the cytotoxicity assays. The protein concentration of the CFE was determined using the Qubit Protein Assay Kit in a Qubit 3.0 fluorometer (Thermo Fisher, Waltham, MA, USA), according to the manufacturer's instructions.

2.3. Proteomic Analysis

The proteomic analysis was performed as reported elsewhere [23], including sample preparation, liquid chromatography–mass spectrometry (HPLC-MS) analysis and protein identification. Briefly, the proteins in the sample were reduced using RapiGest and dithiothreitol (Thermo Fisher), then alkylated using iodoacetamide (Thermo Fisher) and finally digested using LysC-Trypsin and Trypsin (Mass Spec grade, Promega, Madison, WI, USA) in two-steps. After proteolysis was stopped by adding formic acid (Thermo Fisher), the samples were dried and desalted using C18 spin columns (Thermo Fisher). The samples were stored at -20 °C until the analysis. A total nominal amount of 1.5 µg protein was then analyzed from each sample using a Dionex Ultimate 3000 nanoRSLC (Dionex, Sunnyvale, CA, USA) system coupled to a Bruker Maxis II ETD mass spectrometer (Bruker Daltonics GmbH, Bremen, D) via a CaptiveSpray nanobooster ion source. For peptide separation, an ACQUITY UPLC M-Class Peptide BEH C18 column (Waters, Milford, MA, USA) was used. Proteins were identified by searching against the Uniprot Aechinodermata database using the Byonic (v4.2.10, Protein Metrics Inc., Cupertino, CA, USA) software search engine. Protein hits were further filtered by Scaffold (version 4.11, Proteome Software, Inc., Portland, OR, USA). Quantitative proteomic information was provided by Scaffold's quantitative analysis. Subsequently, protein identification was also performed by BlastP comparison to the non-redundant protein sequence database (available at https://blast.ncbi.nlm.nih.gov/Blast.cgi?PAGE=Proteins; accessed on 1 May 2022). An expected value <1 was set as the cutoff.

2.4. Cell Cultures

HepG2 hepatocellular carcinoma cells, taken from laboratory stocks, were grown in D-MEM medium (Sigma, St. Louis, MO, USA) plus 10% fetal calf serum (FCS; Thermo Fisher, Waltham, MA, USA) and antibiotics (100 U/mL penicillin and 100 µg/mL streptomycin; Thermo Fisher, Waltham, MA, USA), at 37 °C in a 5% CO_2 atmosphere.

2.5. Viability Assays

An MTT assay was performed to evaluate cell viability under control and treated conditions [24]. Briefly, after plating at a concentration of 5500 cells/well in 96-well plates and overnight adhesion, HepG2 cells were exposed for 24 or 48 h to the CFE at 2.5, 5, 10 or 15 µg/mL concentrations. After the addition of MTT and the incubation with the solubilization buffer, the absorbance of the solubilized formazan was evaluated in an automated microplate reader at a $\lambda = 550$ nm. After calculation of the viability ratio between treated and control cells, the half maximal inhibitory concentration (IC_{50}) at 24 h ($IC_{50}24$) and 48 h ($IC_{50}48$) was evaluated with the online calculator available at https://www.aatbio.com/tools/ic50-calculator (accessed on 6 March 2021). The subsequent biological assays were performed, exposing cells to either the $IC_{50}24$ or the $IC_{50}48$ of the CFE for 24 or 48 h, respectively.

In parallel experiments, HepG2 cells were co-treated with 1 nM rapamycin (Sigma), a macrolide antibiotic functioning as an autophagy activator [25] and IC_{50} of the CFE for 24 h. The reversion of the cytotoxic effect, if any, was monitored through the trypan blue exclusion test. DMSO-treated cells were used as controls.

2.6. Wound Healing Assay

The scratch/wound healing assay, commonly used to study the cell motile attitude in vitro, was performed according to [23]. Briefly, HepG2 cells were seeded in 6-well plates and allowed to form a sub-confluent monolayer. Then, after the cell layer was scraped three times in parallel with a 200 µL pipette tip and a perpendicular line was drawn with a permanent marker, the culture medium was replaced with either an unsupplemented medium (control sample) or medium containing the CFE at the $IC_{50}24$ (treated sample). Selected sites of intersection between the scratched monolayer and the drawn line were photographed under a phase-contrast microscope at fixed time intervals in the 24 h following the beginning of the assay. The quantification of the wound area was performed using the ImageJ/Fiji® plug-in.

2.7. Flow Cytometry Assays

For each of the following analyses, three independent assays were performed as described in [20,23,26,27] using a FACSCanto flow cytometer (BD Biosciences, Franklin Lakes, NJ, USA) and evaluating 10,000 events. The resulting fcs files were processed with the Floreada analysis tool available at https://floreada.io (accessed on 1 April 2022). In order to exclude debris, which displayed low FSC values, from the analyses, gating in the FSC vs. SSC plot was performed for every determination. Moreover, gating in the FSC-H vs. FSC-A plot was also performed to exclude doublets and multiplets in cell cycle analyses.

2.7.1. Cell Cycle Distribution Assay

Cell cycle distribution was determined by fixation of control and exposed cells with cold 70% ethanol, incubation with 40 µg RNAse A/mL and final staining with 20 µg propidium iodide/mL before the flow cytometrical analysis.

2.7.2. Apoptosis Assay

The percentage of apoptotic cells under control and exposed preparations was evaluated using the Annexin V-FITC kit (Canvax Biotech, Cordoba, Spain), following the manufacturer's instructions.

2.7.3. Analysis of Mitochondrial Transmembrane Potential (MMP)

The MMP state was examined using the dye JC1 (Molecular Probes, Eugene, OR, USA), which exhibits potential-dependent accumulation and a fluorescence emission shift from green (~529 nm) to red (~590 nm) in intact mitochondria whereas, in the case of dissipated MMP, a decrease in the red/green fluorescence intensity ratio occurs. In parallel with the negative controls and CFE-treated samples, HepG2 cells were also exposed to 1 μM valinomycin, an MMP loss-inducer K+ ionophore, as a positive control.

2.7.4. ROS Assay

The extent of intracellular ROS production was evaluated using the ROS Detection Assay Kit (Canvax Biotech, Cordoba, Spain) according to the manufacturer's instructions.

2.7.5. AVO Accumulation Assay

The extent of AVO accumulation was examined by fixing control and treated cells with cold 70% ethanol and staining with 100 μg acridine orange/mL (Sigma) for 20 min in the dark before the flow cytometric analysis.

2.8. Statistics

The analysis of variance (ANOVA) tests was performed with SigmaPlot 11.0 software (SYSTAT, San Jose, CA, USA). A p-value < 0.05 was considered statistically significant.

3. Results

In the first set of assays, we examined the effect exerted by exposure to *A. lixula*'s CFE on HepG2 cells' survival via an MTT assay. As shown in Figure 2, cell exposure for 24 or 48 h to either preparation caused a dose-dependent decrease in cell viability with an average $IC_{50}24$ of 15.7 μg/mL and $IC_{50}48$ of 7.45 μg/mL, respectively. These IC_{50} concentrations were used in the subsequent analyses, which aimed to gather more detailed data on the subcellular aspects of CFE-induced toxicity on HepG2 cells via a panel of flow cytometric assays.

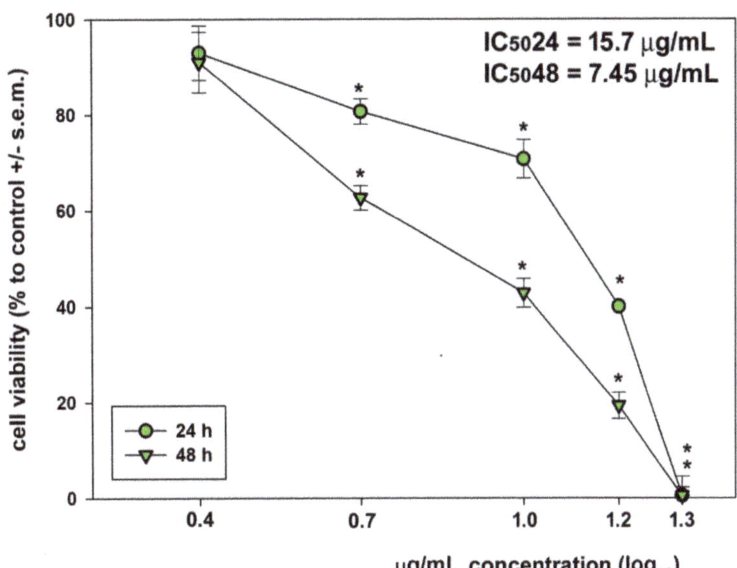

Figure 2. Dose-response effect of the CFE from *A. lixula* at 2.5, 5, 10 and 15 μg/mL concentration on the viability of HepG2 after either 24 or 48 h of exposure. The error bars correspond to the standard error of the mean (s.e.m.) of three independent measurements. * $p < 0.05$.

We initially examined which kind of perturbation of the cell cycle progress could be found after treatment with the preparation. For this purpose, HepG2 cells exposed to the $IC_{50}24$ or $IC_{50}48$ of the CFE were stained with propidium iodide (PI) and submitted to flow cytometric analysis of the distribution of cell population in the cycle phases. As shown in Figure 3, if compared to controls, the result of 24 h of exposure to the extracts was a decrease in the proportion of cells in the G_0/G_1 phase fraction (average control vs. CFE = 47.4% vs. 29.72%), and an increase in the cell percentages in the sub-G_0 fractions (average control vs. CFE = 8.4% vs. 26.18%). No significant change was observed for the other cell cycle phases. Only a progressive increase in the sub-G_0 fraction could be observed at earlier times, i.e., 2 and 4 h (not shown). In addition, cell treatments with the $IC_{50}48$ of the CFE for 48 h determined a generalized decrease in all the cell percentages in the cycle phases and a further increase in the sub-G_0 fraction, thus indicating the generalized exacerbation of the cell cycle arrest and the induction of cell death as results of the longer-term exposure.

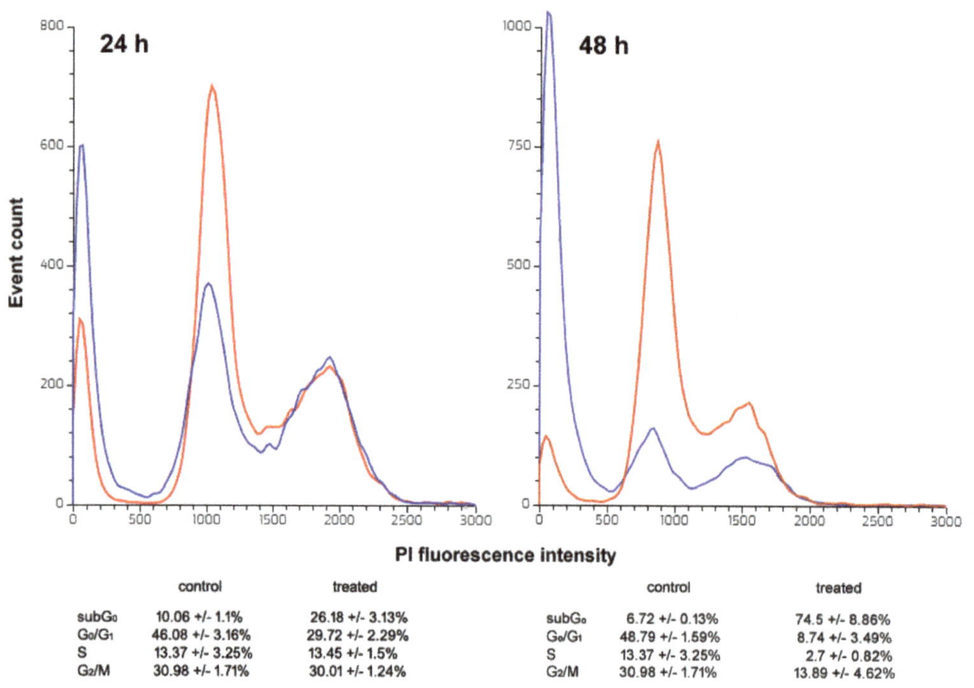

	control	treated		control	treated
subG$_0$	10.06 +/- 1.1%	26.18 +/- 3.13%	subG$_0$	6.72 +/- 0.13%	74.5 +/- 8.86%
G$_0$/G$_1$	46.08 +/- 3.16%	29.72 +/- 2.29%	G$_0$/G$_1$	48.79 +/- 1.59%	8.74 +/- 3.49%
S	13.37 +/- 3.25%	13.45 +/- 1.5%	S	13.37 +/- 3.25%	2.7 +/- 0.82%
G$_2$/M	30.98 +/- 1.71%	30.01 +/- 1.24%	G$_2$/M	30.98 +/- 1.71%	13.89 +/- 4.62%

Figure 3. Representative DNA profiles of control (red line) and CFE-treated (blue line) HepG2 cells after 24 and 48 h of exposure. The cell population distribution in the cycle phases in triplicate experiments is reported in the annexed panels as the mean $\pm$ s.e.m.

Based on the accumulation of HepG2 cells treated with $IC_{50}24$ CFE for 24 h in the sub-G_0 phase fraction, indicative of the occurrence of DNA fragmentation ascribable to the onset of necrosis and/or apoptosis, the tumor cells were submitted to annexin V-FITC/PI staining with the aim to determine the possible externalization of phosphatidylserine, a hallmark of apoptosis promotion. On the one hand, the obtained results, shown in Figure 4, demonstrate that the percentage of the viable annexin-V$^-$/PI$^-$ cells decreased from about 84% of the controls to about 56% after the treatment. On the other hand, the percentage of late apoptotic cells (annexin-V$^+$/PI$^+$) increased from about 14% of the controls to about 41%. This result is in accordance with the previous data regarding the increasing amount of the sub-G_0 cell population already at 24 h of exposure to the CFE. No significant difference was found between the necrotic, i.e., annexin-V$^-$/PI$^+$, cell populations in the two experimental conditions.

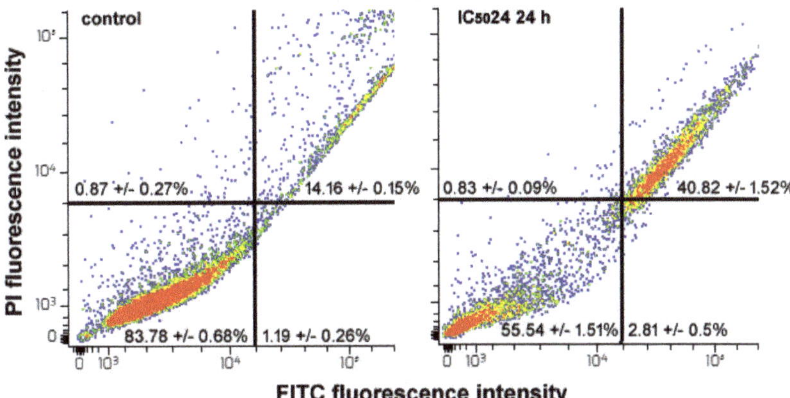

Figure 4. Flow cytometric assays for apoptosis in HepG2 cells cultured under control conditions or exposed to CFE $IC_{50}24$ for 24 h. The plots show the results of representative experiments, and the percentages indicated as the mean $\pm$ s.e.m. of three independent experiments refer to viable annexin-V^-/PI^- cells (bottom left quadrant), early apoptotic annexin-V^+/PI^- cells (bottom right quadrant), late apoptotic annexin-V^+/PI^+ cells (top right quadrant) and necrotic annexin-V^-/PI^+ cells (top left quadrant).

Then, we determined whether cell exposure to the extract could impair the mitochondrial function by monitoring mitochondrial polarization status with the membrane-permeable JC1 dye, which is sensitive to changes in the MMP, thereby evaluating the percentage of cells with bright green/bright red emission (indicative of intact MMP) and of those ones with bright green/dim red emission (indicative of MMP collapse). Figure 5 shows that the CFE determined the dissipation of the MMP. In particular, after 24 h and 48 h of exposure to the preparations at the $IC_{50}24$ and $IC_{50}48$, the dim red-emitting cells were found to increase from about 27% and 25% of the controls to about 48% and 98%, respectively. In the latter case, the value obtained was coincident with that of the positive control treated for 48 h with valinomycin.

The ability of the CFE to dysregulate mitochondrial metabolism was also investigated through the evaluation of the accumulation of ROS. The produced amount of peroxide-like and nitric oxide-derived reactive molecules was monitored by flow cytometric analysis of the oxidation extent of the H_2DCFDA probe to green-emitting DCF at early (4 h of exposure) and late (24 h of exposure) time points. By analogy with other experimental models [23,28–30], two distinct cell subpopulations endowed with low (ROS^-) and high rates (ROS^+) of ROS generation were found in each sample. Thus, we evaluated the mean fluorescence intensity (MFI) of the events associated with the ROS^+ subpopulations in the different experimental conditions to compare their rates. The results obtained show the occurrence of a drastic down-regulation of ROS generation after short-term incubation with the CFE (average control cells' MFI vs. treated cells' MFI = 11,662 vs. 2885), which remained approximately constant after 24 h of exposure (average MFI = 2064) although a decrease in ROS accumulation could be observed at this time point also under control conditions (average MFI = 6710) (Figure 6). As reported by Raja et al. [29], we also evaluated the ratio between ROS^+ and ROS^- cells within the whole population. As shown in Figure 7, CFE treatment correlated with very low levels of basal ROS (average control cells' ratio vs. treated cells' ratio = 1.39 vs. 0.18 at 4 h and 8.73 vs. 0.07 at 24 h). Cumulatively, the obtained data are consistent with the idea that the switching-off of ROS production contributes to the cytotoxic activity of the CFE.

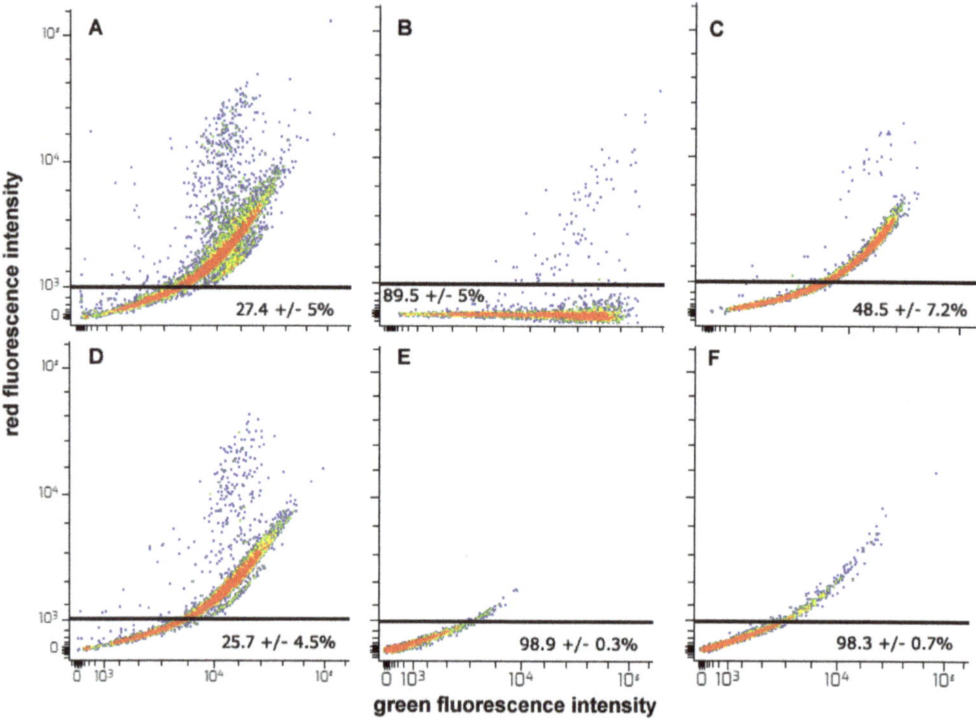

Figure 5. Flow cytometric assays for MMP in HepG2 cells cultured for 24 (**A–C**) or 48 h (**D–F**) under control conditions (**A,D**), in the presence of 1μM valinomycin (**B,E**) and of either CFE IC$_{50}$24 (**C**) or CFE IC$_{50}$48 (**F**). The plots display the outputs of representative experiments, and the percentages in the bottom quadrants of each frame expressed as the mean ± s.e.m. of triplicate experiments are related to dim red-emitting cells that underwent collapse of MMP.

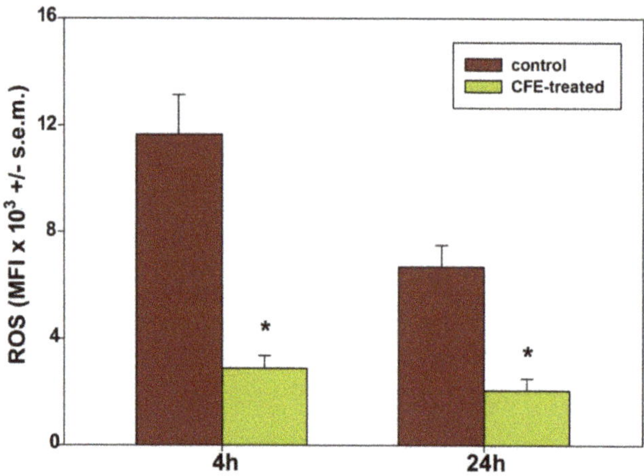

Figure 6. Bar graphs showing the ROS-associated MFI in ROS$^+$ cell subpopulations of control and IC$_{50}$24 CFE-treated HepG2 cells for 4 and 24 h. The error bars correspond to the standard error of the mean (s.e.m.) of three independent measurements. * $p < 0.05$.

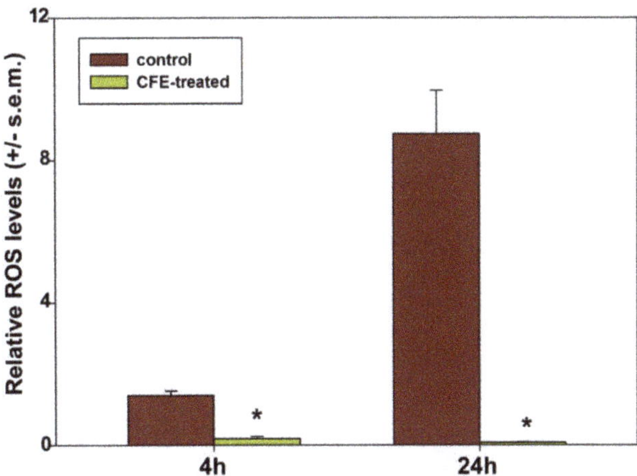

Figure 7. Bar graphs showing the ROS^+ cell/ROS^- cell ratio in control and $IC_{50}24$ CFE-treated HepG2 cells for 4 and 24 h. The error bars correspond to the standard error of the mean (s.e.m.) of three independent measurements. * $p < 0.05$.

In order to determine whether the CFE could affect the autophagic behavior of HepG2 cells, acridine orange staining was performed to label AVOs, a hallmark of autophagic cells that is indicative of autophagosome accumulation and autolysosome formation [31], at early (4 h of exposure) and late (24 h of exposure) time points. As shown in Figure 8, in this case, two distinct cell subpopulations characterized by low (AVO^-) and high rate (AVO^+) of acridine orange fluorescence were found after the flow cytometric assays and plot analyses. The present data confirmed that HepG2 cells are characterized by a high basal level of autophagy, as already indicated in [23,32]. In fact, the percentage of AVO^+ cells accounted for about 90% and 70% at 4 and 24 h of culture under control conditions. Conversely, the percentage of AVO^+ cells decreased after 4 h of treatment with $IC_{50}24$ CFE to about 70%, and the prolongation of the time of exposure to 24 h determined the complete disappearance of the AVO^+ cell fraction, whose proportion dropped to about 0.2% of the total population.

In order to verify to what extent inhibition of the autophagic process could be responsible for the observed CFE-dependent cytotoxicity on HepG2 cells, cells were co-incubated with $IC_{50}24$ CFE and 1 nM rapamycin (sirolimus), a known inhibitor of mTOR (mammalian target of rapamycin) serine/threonine protein kinase, which acts as autophagy promoter. As shown in Figure 9, this co-treatment was unable to reverse the CFE-triggered decrease in cell number, thereby suggesting the occurrence of extensive and widespread damage induced by exposure to CFE.

Lastly, we examined the effect of the incubation with $IC_{50}24$ CFE from A. lixula on HepG2 cells' migratory ability through a scratch wound healing assay. As expected, based on some previously published results [23,33,34], the panel of micrographs in Figure 10 shows that control HepG2 cells exhibited a locomotory attitude that determined the reduction in the denuded area (mean area % = 23 at time 0), which had already started at 2 h from the scratch time (mean area % = 17) and led to its partial obliteration after 6 h (mean area % = 6) and total closure within 22 h. Notably, since the doubling time of HepG2 cells is $\geq$24 h [35,36], the observed effect could not be ascribable to cell proliferation. In contrast, the exposure to the CFE inhibited the cells' ability to migrate into the scratched area and, in addition, starting from 2 h of exposure and in line with the previous data, the cells displayed prominent signs of suffering and damage, such as rounding and clumping, ultimately determining their visible detachment from the substrate.

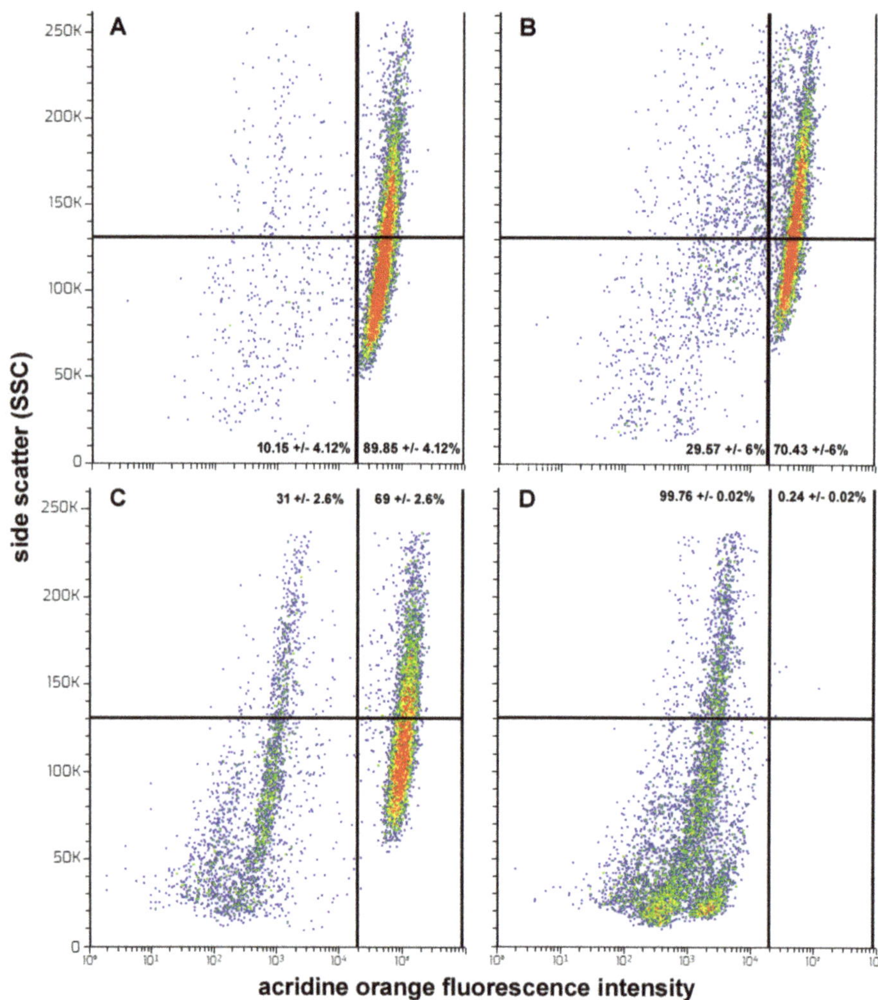

Figure 8. Flow cytometric assays for AVO accumulation in HepG2 cells grown under control conditions for 4 h (**A**) and 24 h (**C**) or incubated with IC_{50}24 CFE for 4 h (**B**) and 24 h (**D**). The plots show the results of representative experiments, and the percentages indicated as the mean $\pm$ s.e.m. of three independent experiments refer to AVO$^+$ cells (right quadrants) and AVO$^-$ cells (left quadrants).

In light of the observed cytotoxic role played by the CFE from A. lixula on HepG2 tumor cells, proteomic analysis of the preparations was performed after proteolysis of the samples and MS sequencing to detect bioactive components. Overall, 1952 obtained spectra matched forward peptides, and the final output reported 104 forward and 20 reverse proteins and 329 unique forward peptides. The estimated spectrum-level FDR on true proteins was 0.3%. A bioinformatic similarity search against the BlastP database identified 28 proteins contained in the CFE and potentially associated with the various aspects related to the derangement of HepG2 cell biological activities reported previously here, as well as with exosome secretion. The peptide sequences and the results of alignments selected on the basis of sorting by the best E value are reported in Table 1.

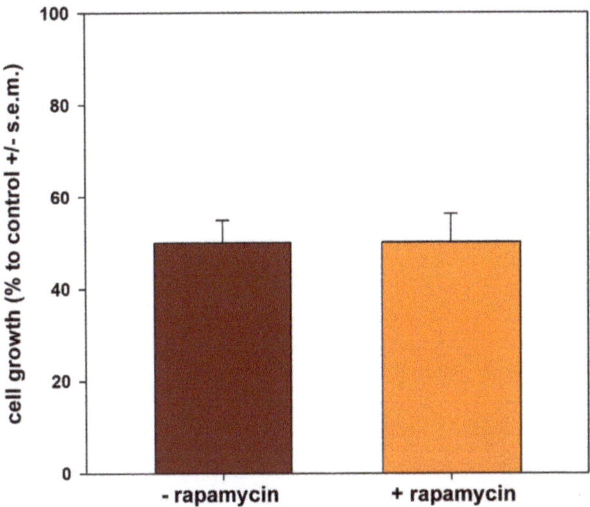

Figure 9. Bar graphs showing the absence of a statistically significant effect by 1 nM rapamycin co-treatment on CFE-induced decrease in cell number after 24 h of exposure. The error bars correspond to the standard error of the mean (s.e.m.) of three independent measurements.

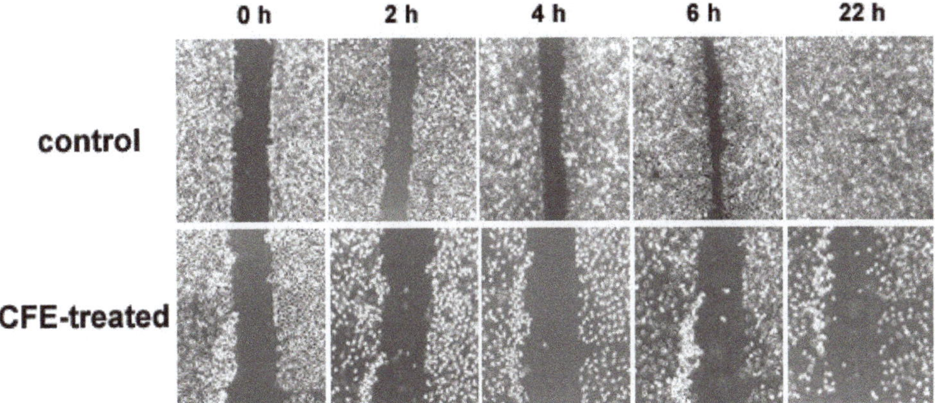

Figure 10. Panel of representative phase-contrast micrographs acquired during the wound healing assays at different time points under control conditions and in the presence of IC_{50} 24 CFE. The assay was performed in triplicate. Microscopical magnification = 20×.

Table 1. Cytotoxic activity-associated protein profile of the CFE from *A. lixula*.

Peptide Sequence(s)	Sequence ID (nr. of Matches/Range)	Expected	Identities (%)	Positives (%)	Protein Description	Organism (Sea Urchin)	Selected GO Annotations
AIGNIGEPVER	BAL43173.1 (1/501–511)	2×10^{-7}	100		Apolipoprotein B-like protein, partial	*Mesocentrotus nudus*	
AAIQSMR	(1/543–549)	0.003	100	100			
LN [+0.984]				100			
QLHYLVGMEPSATDATVFLR	(1/807–828)	5×10^{-19}	100	100			-lipid transport
AYTIDAQYK	(1/1069–1077)	1×10^{-5}	100	100			
LQSPMIK	(1/1815–1821)	0.12	88	88			
QGITFNVFSSYPTLNK	(1/5231–5246)	9×10^{-13}	100	100			
				100	Apolipoprotein B-like protein, partial	*Strongylocentrotus intermedius*	
LALIALK	BAL43173.1 (1/494–500)	0.021	100				
NTLPTKDTIEQEK	AAC26833.1 (1/26–39)	1×10^{-7}	93	100	Thymosin beta	*Strongylocentrotus purpuratus*	-actin monomer binding -negative regulation of cell cycle G_1/S phase transition -regulation of cell migration
RGTIEVSKYDGSYR	XP_041457866.1 (1/572–584)	2×10^{-9}	100	100	Low-density lipoprotein receptor-related protein 1B-like	*Lytechinus variegatus*	-integral component of membrane -receptor-mediated endocytosis
GEQVYGITTYADTR	(1/725–738)	9×10^{-10}	94	93			
GNPTTVFPEIFK	XP_030853362.1 (1/206–217)	1×10^{-8}	100	100	Cryptochrome-2 isoform X1	*Strongylocentrotus purpuratus*	-circadian regulation of gene expression -DNA binding
	XP_030853363.1 (1/79–90)	1×10^{-8}	100	100	Cryptochrome-2 isoform X2		
TSPNSLEPSTTVLSPYLK	XP_030853362.1 (1/367–386)	1×10^{-14}	100	100	Cryptochrome-2 isoform X1		
	XP_030853363.1 (1/79–90)	1×10^{-14}	100	100	Cryptochrome-2 isoform X2		
VYSPIVFGK	XP_030853362.1 (1/540–548)	2×10^{-4}	91	100	Cryptochrome-2 isoform X1		
	XP_030853363.1 (1/413–421)	2×10^{-4}	91	100	Cryptochrome-2 isoform X2		
LAAEGLR							
IGTYGETR	XP_041456451.1 (1/62–68)	0.007	100	100	Arylsulfatase-like	*Lytechinus variegatus*	-sulfuric ester hydrolase activity -lysosome
VFLPWTTTGLPK	XP_041464311.1 (1/109–118)	3×10^{-4}	100	100			
SLSWEGGHR	XP_041460692.1 (1/146–157)	1×10^{-8}	100	100			
	(1/352–360)	1×10^{-4}	91	100			

Table 1. *Cont.*

Peptide Sequence(s)	Sequence ID (nr. of Matches/Range)	Expected	Identities (%)	Positives (%)	Protein Description	Organism (Sea Urchin)	Selected GO Annotations
LVDGSSSNEGR	XP_030831017.1, (20/25–35, 132–142, 239–249, 346–356, 453–463, 560–570, 667–677, 774–784, 881–891, 988–998, 1098–1108, 1353–1363, 1460–1470, 1567–1577, 1674–1684, 1781–1791, 1995–2005, 2102–2112, 2319–2329, 2426–2436)	2×10^{-6}	100	100	Deleted in malignant brain tumors (DMBT) 1 protein isoform X1		
	XP_030831023.1 (20/same as isoform X1 except for 1350–1360, 1457–1467, 1564–1574, 1671–1681, 1778–1788, 1992–2002, 2099–2109, 2316–2326, 2423–2433)	2×10^{-6}	100	100	DMBT1 protein isoform X2		
	XP_030831029.1 (20/same as isoform X2)	2×10^{-6}	100	100	DMBT1 protein isoform X3		
	XP_030831072.1 (19/same as isoform X1 without 2426–2436)	2×10^{-6}	100	100	DMBT1 protein isoform X9		
	XP_030831034.1 (19/same as isoform X9 except for 1888–1898, 2212–2222, 2319–2329)	2×10^{-6}	100	100	DMBT1 protein isoform X4		
	XP_030831054.1 (19/same as isoform X9)	2×10^{-6}	100	100	DMBT1 protein isoform X7		
	XP_030831048.1 (19/same as isoform X4 except for 1995–2005)	2×10^{-6}	100	100	DMBT1 protein isoform X6		
	XP_030831064.1 (19/same as isoform X6 except for 991–1001, 1246–1256)	$2e^{-6}$	100	100	DMBT1 protein isoform X8	*Strongylocentrotus purpuratus*	-membrane -scavenger receptor activity -endocytosis
	XP_030831041.1 (19/same as isoform X9 except for 2212–2222, 2319–2329)	2×10^{-6}	100	100	DMBT1 protein isoform X5		
	XP_030831078.1 (19/same as isoform X1 except for 1202–1212, 1309–1319, 1416–1426, 1523–1533, 1630–1640, 1844–1854, 1951–1961, 2166–2176, 2275–2285)	2×10^{-6}	100	100	DMBT1 protein isoform X10		
	XP_030831079.1 (19/same as isoform X9 except for 2209–2219)	2×10^{-6}	100	100	DMBT1 protein isoform X11		
	XP_030831086.1 (17/25–35, 132–142, 239–249, 346–356, 453–463, 560–570, 667–677, 774–784, 988–998, 1095–1105, 1202–1212, 1309–1319, 1416–1426, 1630–1640, 1737–1747, 1954–1964, 2061–2071)	2×10^{-6}	100	100	DMBT1 protein isoform X12		
MLGYNGALSAPR	XP_030834718.1 (1/278–289)	5×10^{-6}	100	100			
LVGGSGPHEGR	XP_030855705.1 (1/677–687)	0.27	69	76	DMBT1 protein-like		
	(1/1259–1269)	1×10^{-6}	100	100	DMBT1 protein		
	(1/1917–1927)	0.27	77	76			
	(1/2449–2459)	0.017	77	76			
	(1/2129–2139)	0.096	77	84			
	(1/1590–1600)	0.19	77	76			
	(1/2553–2563)	0.78	69	69			
	XP_030851619.1 (1/839–849)	0.49	85	100			
LVGGSNDADGR	(1/1443–1453)	0.061	92	100	DMBT1 protein-like		

Table 1. *Cont.*

Peptide Sequence(s)	Sequence ID (nr. of Matches/Range)	Expected	Identities (%)	Positives (%)	Protein Description	Organism (Sea Urchin)	Selected GO Annotations
VLEQSDLVVDNEYALYTLIQPK	XP_030847432.1 (1/183–192)	0.092	75	83	BTB/POZ domain-containing protein At2g30600	*Strongylocentrotus purpuratus*	-protein ubiquitination
MIFIQGFVPAGADR	CBX45521.1 (1/37–50)	3×10^{-11}	100	100	Galectin-8 protein	*Paracentrotus lividus*	-membrane -integrin binding
NLLSVAYK	XP_780530.1 (1/20–26)	0.35	100	100	14-3-3-like protein 2	*Strongylocentrotus purpuratus*	-signal transduction
RLVELINQ	XP_030850878.1 (1/715–722)	1×10^{-4}	100	100	Centrosomal protein of 85 kDa (CEP85) isoform X1	*Strongylocentrotus purpuratus*	-spindle pole -regulation of mitotic centrosome separation
VLAYNVFELR	XP_041469541.1 (1/103–112)	4×10^{-7}	100	100	Sphingomyelinase C 2-like	*Lytechinus variegatus*	- acid sphingomyelin phosphodiesterase activity
GDSTLQFKPK	XP_041482277.1 (1/359–368)	5×10^{-6}	100	100	Tripartite motif-containing protein 45-like	*Lytechinus variegatus*	-ubiquitin-protein transferase activity -zinc ion binding
RLITRNSYE KEEELES RAAQFAALM KHQSYNVAMEWANIV	XP_030831990.1 (1/1592–1600) (1/2501–2507) (1/2775–2783) (1/3026–3040)	4×10^{-6} 0.003 2×10^{-5} 2×10^{-13}	100 100	100 100	Chromodomain-helicase-DNA-binding protein 8	*Strongylocentrotus purpuratus*	binding -negative regulation of canonical Wnt signaling pathway
DN [+0.984] IVIGGQAGVYDPNR	XP_001198520.2 (1/164–179)	1×10^{-12}	100	100	Hemicentin-2	*Strongylocentrotus purpuratus*	-calcium ion binding -response to stimulus
RVASGPLGLI	XP_030854745.1 (1/1903–1912)	1×10^{-5}	100	1000	WD repeat-containing protein 81	*Strongylocentrotus purpuratus*	-cytoplasmic vesicle
QINLDLLR	XP_041482078.1 (1/665–674)	2×10^{-4}	100	100	TBC1 domain family member 2B-like isoform X1	*Lytechinus variegatus*	-activation of GTPase activity
YPHTQLISQMDR	XP_030846390.1 (1/372–383)	1×10^{-9}	100	100	Growth/differentiation factor 8-like	*Strongylocentrotus purpuratus*	-signaling receptor binding -negative regulation of skeletal muscle satellite cell proliferation
MQVQ [+0.984] SGVPK	XP_003726849.3 (1/1–9)	1×10^{-4}	100	100	Selenide, water dikinase	*Strongylocentrotus purpuratus*	-selenocysteine biosynthetic process
KYSEPTVEDISPVEHVE	XP_030841466.1 (1/132–148)	3×10^{-10}	100	100	Kv channel-interacting protein 4-like	*Strongylocentrotus purpuratus*	-potassium channel regulator activity -calcium ion binding
DFIVIQNFITEEEEDSLLK	XP_011661956.1 (1/144–162)	2×10^{-16}	100	100	Alpha-ketoglutarate-dependent dioxygenase alkB homolog 7, mitochondrial	*Strongylocentrotus purpuratus*	-mitochondrial matrix -regulation of mitochondrial membrane permeability involved in programmed necrotic cell death

Table 1. *Cont.*

Peptide Sequence(s)	Sequence ID (nr. of Matches/Range)	Expected	Identities (%)	Positives (%)	Protein Description	Organism (Sea Urchin)	Selected GO Annotations
REVPVASL	XP_011668608.2 (1/721–728)	0.001	100	100	Sterile alpha motif domain-containing protein 9-like	*Strongylocentrotus purpuratus*	-intracellular membrane-bounded organelle
LAKALQLSE	XP_030841717.1 (1/221–229)	0.003	91	90	Hematopoietically expressed homeobox protein HHEX homolog	*Strongylocentrotus purpuratus*	-Wnt signaling pathway -negative regulation of cyclin-dependent protein serine/threonine kinase activity
SQIPTK	XP_041485363.1 (1/1915–1920) XP_041485364.1 (1/1424–1431)	0.024 0.024	100 100	100 100	Methylcytosine dioxygenase TET3-like isoform X1 Methylcytosine dioxygenase TET3-like isoform X2	*Lytechinus variegatus*	-methylcytosine dioxygenase activity -chromatin organization
QVTGTGATGR GGSSAQAIR	XP_796458.1 (1/107–116) NP_999723.1 (1/36–44)	2×10^{-5} 4×10^{-4}	100 100	100 100	Histone H1, gonadal-like Histone H1-beta, late embryonic	*Strongylocentrotus purpuratus*	
HLQLAIR NDEELNKLLGGVTIAQGGVLPNIQAVLLPK KLLSGVTIAQGGVLPNIQAVLLPK GDEELDSLIK	XP_030843320.1 (1/83–89) XP_001175793.2 (1/89–118) NP_999718.1 (1/97–119) NP_001116980.1 (1/93–102)	0.002 6×10^{-26} 3×10^{-18} 7×10^{-6}	100 100 100 100	100 100 100 100	Histone H2A-like, partial Histone H2A, embryonic-like Late histone H2A.L3 Histone H2A.V	*Strongylocentrotus purpuratus*	
RKESYGIYTYKV VLK QVHPDTGISSR AMSIMNSFVNDVFER STITSR IAAEASR RLLLPGELAKH HAVSEGTK YRPGTVALR STELLIR EIAQDFKTELR FQSSAVMALQEASEAYLVGLFEDTN [+0.984] LC [+57.021] AIH-AK	P022892 (1/33–42) (1/43–45) (1/46–56) NP_001229615.1 (1/56–70) XP_030836142.1 (1/84–89) AAB48832.1 (1/75–81) (102–110) (110–118) XP_030830328.1 (1/42–50) (1/58–64) (1/74–84) (1/85–112)	0.002 6×10^{-12} 0.10 0.016 6×10^{-5} 0.002 2×10^{-5} 0.008 7×10^{-8} 3×10^{-22}	100 100 100 100 100 100 100 100 100 94	100 100 100 100 100 100 100 100 100 93	Histone H2B, embryonic Histone H2B-like Histone H2B Cleavage stage histone H2B Histone H3, embryonic-like	*Strongylocentrotus purpuratus* *Psammechinus miliaris* *Strongylocentrotus purpuratus*	
DNIQGITKPAIR RISGLIYEETR VFLENVIR	CAA75404.1 (1/1–12) (1/22–32) (1/37–44)	2×10^{-7} 3×10^{-6} 2×10^{-4}	100	100	Histone H4, partial	*Arbacia lixula*	
TLYGFGG	XP_011664585.2 (1/97–102)	0.051	100	100	Histone H4-like	*Strongylocentrotus purpuratus*	

4. Discussion

Sea urchins' coelomic fluid represents a pseudo-vascular system designed for the transport of nutrients and gases and the fulfillment of the immune defense and stress response reactions. It contains a complex mixture of soluble molecules secreted constitutively by different parts of the echinoderm's body, whose identification and biological characterization are still very limited. In terms of possible biomedical translation, the available literature data only indicated that the cell-free CF from Tripneustes depressus possessed anti-viral properties against Suid herpesvirus type I and rabies virus, whereas that of Echinometra mathaei was endowed with a high radical-scavenging activity [37,38]. Additional studies focused on the characterization of protein components in the CFEs from different echinoderm species, pointing them out as potential sources of anti-microbial and anti-cancer agents [19,39]. Interestingly, D'Alessio et al. [40] recently found the presence of exosome-like extracellular vesicles in Strongylocentrotus purpuratus's CF and characterized their associated protein cargo, thus suggesting an increasing complexity in the composition of the fluid.

Previous results had revealed for the first time the beneficial effects of A. lixula's CFE against human triple-negative breast cancer, using MDA-MB231 cells as a model system [20]. In this case, the biological characterization of the cytotoxic activity highlighted the arrest of the cells in the S phase of the cell cycle, the depolarization of the mitochondria and the up-regulation of ROS production, although no occurrence of apoptotic death and of modification in AVOs' accumulation rate was observed.

Here we focused our attention on determining the possible protective effects exerted by the extract against the HepG2 cell line selected as an in vitro model of cancer of the digestive system, i.e., hepatocellular carcinoma. This aggressive neoplastic histotype represents the most common form of primary liver cancer, ranking sixth in incidence and fourth in mortality among all tumors [41]. Although hepatocarcinogenesis is multifactorial, the main risk factors include chronic fibrosing liver disease and cirrhosis [42]. Within the altered microenvironment of chronic liver injury, the damaged hepatocytes are addressed to a sustained regeneration program coupled with liver progenitor cell expansion, which ultimately results in the accumulation of genetic alterations and progressive cell de-differentiation likely responsible for their neoplastic transformation [43]. The advances in the knowledge of the major molecular signaling pathways involved in the development of hepatocellular carcinoma (HCC) prompted the analysis of the potential preventive and therapeutic roles of natural products, able to intervene in the modulation of the oncogenesis-linked intracellular events [44,45]. However, an accurate biological characterization of the molecular mechanism of action of the agents under study is the essential foundation for the successful development of targeted treatments.

Our data depicted an anti-HCC role of the CFE preparation, at least under the conditions used, but displaying different characteristics from those reported for breast carcinoma cells. In fact, exposure of HepG2 cells to the extract determined cell addressing to the sub-G_0 phase after arrest at the G_2/M phase associated with apoptotic death that reached massive completion after 48 h. During this time, cells underwent gradual mitochondrial depolarization and ROS and AVO depletion. Consistent with the derangement of cells' healthy state, an early block of cells' motile activities was also demonstrated, thus suggesting that the CFE may also act as a potential suppressor of HCC metastatic ability. The panel of study endpoints chosen to test the efficacy of the CFE on HCC cells provides a framework of phenotypic changes that converge to confirm its powerful inhibitory effect.

As a complement to the study of the effect of CFE treatment on cell phenotype, we undertook a search at the proteomic level of CFE's molecular constituents that may conceivably be involved in the observed cytotoxic activity. This study predicted a number of putative anti-cancer proteins responsible for the lethal effect on HepG2 cells. In addition, as already reported for Holothuria tubulosa's CFE [23], histones, which can be considered exosome-linked signatures, were also detected, suggesting the release of extracellular vesicles carrying, among the others, the intracellular constituents likely causing the anti-

HepG2 effects. It is known that extracellular histones are involved in exosome-mediated adhesion and could conceivably be involved in exosome uptake by target cells and the subsequent activation of signaling [46]. Exosomes are conceivably preserved intact by the method of preparation of the CFE and, therefore, could be ready to fuse with and transfer their cargo into cancer cells. In addition, to strengthen our hypothesis, Tadayoni Nia et al. [47] reported that the siRNA-mediated silencing of the gene coding for the transmembrane WD repeat-containing protein 81, one of the protein signatures found in our analysis, was able to reduce the release of exosomes by human glioblastoma cells.

Among the protein signatures identified, most of them may be mainly, but not exclusively, associated with the activation of programmed cell death. Lipid-free apolipoprotein B has been shown to exert cytotoxic effects on several cytotypes through binding and disturbing the structure of their plasmalemma, thereby triggering apoptosis [48]. The early report by Hall [49] ascribed to thymosin β10 an apoptosis-controlling role by functioning as an actin-mediated tumor suppressor, which induces cytoskeletal disorganization. In addition, Thr^{20} was found to be specifically required for actin sequestration and apoptosis promotion in ovarian cancer cells, whose thymosin-dependent inhibition of the Ras pathway was acknowledged [50,51]. Arylsulfatases are common components of toxic secretions of different animal organisms [52]; of note, Lai et al. [53] reported the effect of the up-regulation of SULF1, coding for arylsulfatase-1, on the decrease in the proliferation rate and motile attitude in vitro and tumorigenesis in vivo of liver tumor cell lines, and the potentiation of cell sensitivity to apoptotic stimuli possibly via inhibition of both the MAP kinase/ERK and PI3 kinase/AKT kinase pathways. Deleted In Malignant Brain Tumors 1 (DMBT1), which was also found in the CFE of Holothuria tubulosa [23], acts as a tumor growth suppressor and an effector of genetic resistance to liver carcinogenesis in rats and humans [54]. Its overexpression in GBC-SD gallbladder cancer cells was proven to decrease cell proliferation rate and induce apoptosis via stabilization of the phosphatase and tensin homolog (PTEN) and inhibition of the PI3K-Akt pathway, also reducing xenograft tumor growth in vivo [55]. BTB/POZ domain-containing proteins were found to inhibit proliferation, locomotion and invasion; impair cell cycle; and promote apoptosis in a panel of different tumor cell lines, also restraining xenograft tumor growth [56–58]. Hadari et al. [59] demonstrated that the integrin-binding protein galectin-8 was responsible for the inhibition of the adhesion of human lung cancer cells to the substrate, thereby addressing them to apoptosis. Autelli et al. [60] produced evidence that the increase of intracellular ceramide generated by acid sphingomyelinase-mediated sphingomyelin cleavage determined the activation of caspases and the stimulation of the intrinsic pathway of apoptosis in rat hepatoma cells co-treated with TNF and cycloheximide. Tripartite motif-containing protein 45 is a tumor suppressor member of the RING-finger-containing E3 ligases. Data from Peng et al. [61] and Zhang et al. [62] demonstrated its ability to promote cell apoptosis by activating the p38 signal and inhibit proliferation by down-regulating p-ERK in lung cancer cells, as well as to suppress glioblastoma proliferation and tumorigenicity by protecting p53 from degradation and inactivation, thereby allowing cells to undergo apoptosis. The inhibition of the intrinsic pathway of apoptosis was a result of the absence of functioning growth/differentiation factor 8, a.k.a. myostatin, in knockout C2C12 cells; the same factor was also proven to reduce the viability and disrupt the migratory and proliferative potential of MCF-7 breast cancer cells [63,64]. Ultimately, sterile alpha motif domain-containing protein 9 was reported to play a fundamental role in the interferon β-induced apoptotic pathway in human glioblastoma cells [65].

In addition, some protein signatures may be mainly associated with the impairment of the cell cycle and the inhibition of cell proliferation. This is the case of 14-3-3 protein, CEP85, cryptochrome-2, chromodomain-helicase-DNA-binding protein 8, HHEX, low-density lipoprotein receptor-related protein 1B and TET3 [66–74]. In particular, the p53-inducible oncosuppressor 14-3-3 is known to prevent the import of the cyclin B1/cdc2 complex in the nucleus and to bind to CDK2 and CDK4, thus likely affecting the G_1/S transition. Overexpression of CEP85 was found responsible for preventing centrosome disjunction during mitosis. Knockdown of cryptochrome-2, a core component of the circadian clock, was proven to up-regulate osteosarcoma cell proliferation and migration by inducing cell cycle progression and promoting the MAPK and Wnt/β-catenin signaling pathways; similarly, chromodomain-helicase-DNA-binding protein 8 also functions as a tumor suppressor by regulating Wnt/β-catenin signaling and the cell cycle. Overexpressed HHEX was reported to restrain hyperproliferation, metabolism activity, cell size and transformation characteristic of c-Myc tumorigenic activities by binding and breaking c-Myc/Max heterodimers; in particular, dealing with liver cancer, HHEX proved capable of inhibiting tumorigenicity of Hepa1-6 hepatoma cells by suppressing their growth and colony-forming ability in vitro and tumor development in nude mice. Overexpression of full-length LRP1B, coding for the low-density lipoprotein receptor-related protein 1B, led to an impaired proliferation of human non-small cell lung cancer cells, whereas its knockdown exerted the opposite effect. In addition, this protein was found to function as a tumor suppressor also against renal cell cancer, its down-regulation leading to the increase in cell migration and invasion, possibly via actin cytoskeleton remodeling regulated through the Cdc42/RhoA pathway and expressional alteration of the components of focal adhesion complexes. Ultimately, a knockdown of the expression of TET3 was reported to promote the proliferation of HepG2 cells.

One of the identified protein signatures, i.e., alkB homolog 7, could be associated with mitochondrial damage. In fact, data from Fu et al. [75] demonstrated its involvement in the massive loss of mitochondrial homeostasis leading to permeability transition pore opening, NAD^+ efflux into the cytoplasm and, eventually, energy depletion and cellular demise.

Regarding the four remaining signatures listed, it may be inferred that: (i) selenide and water dikinase determine the decrease in ROS accumulation since Na et al. [76] reported that deficient cells showed the down-regulation of genes involved in ROS scavenging; (ii) Kv channel-interacting protein 4 acts as a putative tumor suppressor, as shown in hepatocellular carcinoma [77]; (iii) hemicentin-1 acts as putative cytotoxicity-associated factor, as shown in breast cancer [78]; and, (iv) TBC1 domain family member 2B, a Rab22-binding protein, is an inhibitor of cell motile attitude by blocking E-cadherin degradation necessary for epithelial-mesenchymal transition, as demonstrated with human lung cancer cells [79].

By taking the obtained results and the literature data into consideration, the findings reported here allow us to make the following comments:

(1) For HepG2 cells, anti-cancer agent-triggered growth arrest at the G_2/M phase is an event commonly leading up to apoptosis. This impairment was ascribed to different causes, such as the inhibition of cyclin-dependent kinase-2A, the activation of p38 MAPK signalization or the inhibition of tubulin polymerization, determining the failure of the assembly and dynamics of mitotic spindle structure [80–82]. Therefore, the molecular basis of CFE-stimulated cell cycle block is an interesting aspect that remains to be clarified and can bring to light new intracellular targets for anti-HCC treatment options;

(2) It is known that HepG2 cells are endowed with elevated levels of basal autophagy [23,56]. This allows their survival, active growth and locomotion by fulfilling the related high metabolic and energetic demands. Thus, it is conceivable that the suppression of "protective" autophagy may contribute to CFE's cytotoxic activity. Interestingly, in line with [83], autophagy down-regulation may lead to the inhibition of cell proliferation and the onset of apoptosis via the intrinsic pathway, as suggested in our model system by the derangement of mitochondrial function. On the other hand, the failure of rapamycin co-treatment to restore cell viability suggests that autophagy inhibition is secondary to the other CFE exposure-linked aspects of cell damage, i.e., mitochondrial dysfunction, ROS depletion and apoptosis promotion, the latter determining a massive impairment of cell functions that cannot be rescued by the sole autophagy reactivation;

(3) It is worth mentioning that the HepG2 cell line is characterized by low levels of expression of the cytochrome P450 family 2 subfamily E member 1 (CYP2E1) protein,

a ROS-generating enzyme of the endoplasmic reticulum, and therefore is considered a helpful model to test the formation of ROS mainly from mitochondrial sources [84]. Our results show a drastic reduction in the generation of mitochondrial ROS in exposed cells. It is known that low levels of ROS act as redox-active signaling messengers necessary for cell proliferation and functions and that cancer cells require a greater supply of these molecules; otherwise, they become unable to grow normally [85]. Thus, by analogy with the literature data, it is conceivable that the redox imbalance occurring in CFE-treated HepG2 cells may also contribute to the cytotoxicity-leading vicious cycle by promoting mitochondrial dysfunction and triggering mitochondria-mediated apoptosis.

A limitation of this study is represented by the lack of identification of the "active constituent(s)" of the CFE responsible for the death-promoting activity. Our results indicate that this (these) compound(s) withstand(s) lyophilization, resuspension and freeze–thawing cycles. Nevertheless, the proteomic profile analysis revealed the presence of a rich mix of proteins that can seemingly play anti-cancer roles at different levels. Moreover, molecular docking analyses by Wydananda et al. [39] also demonstrated the potential of peptides from A. lixula's CF to act as anti-non small cell lung cancer agents due to their ability to inhibit different signaling pathways implicated in the progression of this tumoral histotype. Thus, cumulative evidence prompts focusing attention on the protein components of the CFE for a future and more detailed examination of their biological properties and possible applications.

5. Conclusions

Overall, our results provide evidence for the cytotoxic potential of A. lixula's CFE against HCC cells, which comes in addition to the anti-triple negative breast cancer effect previously published [Luparello], thus increasing the interest in the biomedical implementation of the preparation. Future work will focus on the isolation of the substance(s) responsible for the cytotoxic effect and on the more detailed characterization of the underlying molecular mechanism(s) that mediate(s) cell death. Given the great need for developing alternative treatment options against liver cancer, our results indicate that the aqueous extract of A. lixula's CF can be taken into consideration as a potential anti-HCC agent for the development of novel prevention and/or treatment agents, which could also be included as a supplement in functional food or during food processing and packaging.

Author Contributions: Conceptualization, C.L., V.A. and M.V.; investigation, R.B., G.A., V.L., S.S. and M.M.; data curation, C.L., S.S., V.D.S. and M.V.; writing—original draft preparation, C.L. and M.V.; writing—review and editing, C.L. and M.V.; supervision, C.L., V.A. and M.V.; funding acquisition, C.L. and M.V. All authors have read and agreed to the published version of the manuscript.

Funding: This research was funded by The University of Palermo (Italy), grant Fondo Finalizzato alla Ricerca (FFR) 2021, to C.L. and M.V.

Institutional Review Board Statement: The "Assessorato regionale dell'agricoltura, sviluppo rurale e pesca mediterranea" of Sicily has approved the use of A. lixula in this study (authorization nr. 87468 of 25 October 2021).

Informed Consent Statement: Not applicable.

Data Availability Statement: The data presented in the current study are available from the corresponding author upon request.

Conflicts of Interest: The authors declare no conflict of interest.

References

1. Li, C.Q.; Ma, Q.Y.; Gao, X.Z.; Wang, X.; Zhang, B.L. Research Progress in Anti-Inflammatory Bioactive Substances Derived from Marine Microorganisms, Sponges, Algae, and Corals. *Mar. Drugs* **2021**, *19*, 572. [CrossRef] [PubMed]
2. Lindequist, U. Marine-Derived Pharmaceuticals? *Challenges and Opportunities. Biomol. Therap.* **2016**, *24*, 561–571. [CrossRef]

3. Liu, L.; Zheng, Y.Y.; Shao, C.L.; Wang, C.Y. Metabolites from marine invertebrates and their symbiotic microorganisms: Molecular diversity discovery, mining, and application. *Mar. Life Sci. Technol.* **2019**, *1*, 60–94. [CrossRef]

4. Lazzara, V.; Arizza, V.; Luparello, C.; Mauro, M.; Vazzana, M. Bright Spots in The Darkness of Cancer: A Review of Starfishes-Derived Compounds and Their Anti-Tumor Action. *Mar. Drugs* **2019**, *17*, 617. [CrossRef] [PubMed]

5. Luparello, C.; Mauro, M.; Lazzara, V.; Vazzana, M. Collective Locomotion of Human Cells, Wound Healing and Their Control by Extracts and Isolated Compounds from Marine Invertebrates. *Molecules* **2020**, *25*, 2471. [CrossRef]

6. Luparello, C.; Mauro, M.; Arizza, V.; Vazzana, M. Histone Deacetylase Inhibitors from Marine Invertebrates. *Biology* **2020**, *9*, 429. [CrossRef]

7. Mauro, M.; Lazzara, V.; Punginelli, D.; Arizza, V.; Vazzana, M. Antitumoral compounds from vertebrate sister group: A review of Mediterranean ascidians. *Dev. Comp. Immunol.* **2020**, *108*, 103669. [CrossRef]

8. Luparello, C. Marine Animal-Derived Compounds and Autophagy Modulation in Breast Cancer Cells. *Foundations* **2021**, *1*, 3–20. [CrossRef]

9. Šimat, V.; Elabed, N.; Kulawik, P.; Ceylan, Z.; Jamroz, E.; Yazgan, H.; Čagalj, M.; Regenstein, J.M.; Özogul, F. Recent Advances in Marine-Based Nutraceuticals and Their Health Benefits. *Mar. Drugs* **2020**, *18*, 627. [CrossRef]

10. Debeaufort, F. Active biopackaging produced from by-products and waste from food and marine industries. *FEBS Open Bio* **2021**, *11*, 984–998. [CrossRef]

11. Lobine, D.; Rengasamy, K.R.R.; Mahomoodally, M.F. Functional foods and bioactive ingredients harnessed from the ocean: Current status and future perspectives. *Crit. Rev. Food Sci. Nutr.* **2021**, *16*, 5794–5823. [CrossRef] [PubMed]

12. Wangensteen, O.S.; Turon, X.; Pérez-Portela, R.; Palacín, C. Natural or naturalized? Phylogeography suggests that the abundant sea urchin Arbacia lixula is a recent colonizer of the Mediterranean. *PLoS ONE* **2012**, *7*, e45067. [CrossRef] [PubMed]

13. Cakal Arslan, O.; Parlak, H. Embryotoxic effects of nonylphenol and octylphenol in sea urchin Arbacia lixula. *Ecotoxicol.* **2007**, *6*, 439–444. [CrossRef] [PubMed]

14. Carballeira, C.; De Orte, M.R.; Viana, I.G.; Delvalls, T.A.; Carballeira, A. Assessing the toxicity of chemical compounds associated with land-based marine fish farms: The sea urchin embryo bioassay with Paracentrotus lividus and Arbacia lixula. *Arch. Environ. Contam. Toxicol.* **2012**, *2*, 249–261. [CrossRef] [PubMed]

15. Vazzana, M.; Mauro, M.; Ceraulo, M.; Dioguardi, M.; Papale, E.; Mazzola, S.; Arizza, V.; Beltrame, F.; Inguglia, L.; Buscaino, G. Underwater high frequency noise: Biological responses in sea urchin Arbacia lixula (Linnaeus, 1758). *Comp. Biochem. Physiol. A Mol. Integr. Physiol.* **2020**, *242*, 110650. [CrossRef]

16. Cirino, P.; Brunet, C.; Ciaravolo, M.; Galasso, C.; Musco, L.; Vega Fernández, T.; Sansone, C.; Toscano, A. The Sea Urchin Arbacia lixula: A Novel Natural Source of Astaxanthin. *Mar. Drugs* **2017**, *15*, 187. [CrossRef]

17. Galasso, C.; Orefice, I.; Toscano, A.; Vega Fernández, T.; Musco, L.; Brunet, C.; Sansone, C.; Cirino, P. Food Modulation Controls Astaxanthin Accumulation in Eggs of the Sea Urchin Arbacia lixula. *Mar. Drugs* **2018**, *16*, 186. [CrossRef]

18. Stabili, L.; Acquaviva, M.I.; Cavallo, R.A.; Gerardi, C.; Narracci, M.; Pagliara, P. Screening of Three Echinoderm Species as New Opportunity for Drug Discovery: Their Bioactivities and Antimicrobial Properties. *Evid. Based Complement. Alternat. Med.* **2018**, *2018*, 7891748. [CrossRef]

19. Sciani, J.M.; Emerenciano, A.K.; Cunha da Silva, J.R.; Pimenta, D.C. Initial peptidomic profiling of Brazilian sea urchins: Arbacia lixula, Lytechinus variegatus and Echinometra lucunter. *J. Venom Anim. Toxins Incl. Trop. Dis.* **2016**, *22*, 17. [CrossRef]

20. Luparello, C.; Ragona, D.; Asaro, D.M.L.; Lazzara, V.; Affranchi, F.; Arizza, V.; Vazzana, M. Cell-Free Coelomic Fluid Extracts of the Sea Urchin Arbacia lixula Impair Mitochondrial Potential and Cell Cycle Distribution and Stimulate Reactive Oxygen Species Production and Autophagic Activity in Triple-Negative MDA-MB231 Breast Cancer Cells. *J. Mar. Sci. Eng.* **2020**, *8*, 261. [CrossRef]

21. Donato, M.T.; Tolosa, L.; Gómez-Lechón, M.J. Culture and Functional Characterization of Human Hepatoma HepG2 Cells. *Methods Mol. Biol.* **2015**, *1250*, 77–93. [CrossRef] [PubMed]

22. European Medicines Agency. Comparability of Biotechnological/Biological Products Subject to Changes in Their Manufacturing Process. Available online: https://www.ema.europa.eu/en/documents/scientific-guideline/ich-q-5-e-comparability-biotechnological/biological-products-step-5_en.pdf (accessed on 20 November 2021).

23. Luparello, C.; Branni, R.; Abruscato, G.; Lazzara, V.; Drahos, L.; Arizza, V.; Mauro, M.; Di Stefano, V.; Vazzana, M. Cytotoxic capability and the associated proteomic profile of cell-free coelomic fluid extracts from the edible sea cucumber Holothuria tubulosa on HepG2 liver cancer cells. *EXCLI J.* **2022**, *21*, 722–743. [CrossRef] [PubMed]

24. Longo, A.; Librizzi, M.; Chuckowree, I.S.; Baltus, C.B.; Spencer, J.; Luparello, C. Cytotoxicity of the urokinase-plasminogen activator inhibitor carbamimidothioic acid (4-boronophenyl) methyl ester hydrobromide (BC-11) on triple-negative MDA-MB231 breast cancer cells. *Molecules* **2015**, *20*, 9879–9889. [CrossRef]

25. Librizzi, M.; Longo, A.; Chiarelli, R.; Amin, J.; Spencer, J.; Luparello, C. Cytotoxic effects of Jay Amin hydroxamic acid (JAHA), a ferrocene-based class I histone deacetylase inhibitor, on triple-negative MDA-MB231 breast cancer cells. *Chem. Res. Toxicol.* **2012**, *25*, 2608–2616. [CrossRef] [PubMed]

26. Luparello, C.; Ragona, D.; Asaro, D.M.L.; Lazzara, V.; Affranchi, F.; Celi, M.; Arizza, V.; Vazzana, M. Cytotoxic Potential of the Coelomic Fluid Extracted from the Sea Cucumber Holothuria tubulosa against Triple-Negative MDA-MB231 Breast Cancer Cells. *Biology* **2019**, *8*, 76. [CrossRef]

27. Luparello, C.; Asaro, D.M.L.; Cruciata, I.; Hassell-Hart, S.; Sansook, S.; Spencer, J.; Caradonna, F. Cytotoxic activity of the histone deacetylase 3-selective inhibitor Pojamide on MDA-MB-231 triple-negative breast cancer cells. *Int. J. Mol. Sci.* **2019**, *20*, 804. [CrossRef]

28. Belyaeva, E.A.; Dymkowska, D.; Wieckowski, M.R.; Wojtczak, L. Reactive oxygen species produced by the mitochondrial respiratory chain are involved in Cd2+-induced injury of rat ascites hepatoma AS-30D cells. *Biochim. Biophys. Acta* **2006**, *1757*, 1568–1574. [CrossRef]

29. Raja, S.M.; Clubb, R.J.; Ortega-Cava, C.; Williams, S.H.; Bailey, T.A.; Duan, L.; Zhao, X.; Reddi, A.L.; Nyong, A.M.; Natarajan, A.; et al. Anticancer activity of Celastrol in combination with ErbB2-targeted therapeutics for treatment of ErbB2-overexpressing breast cancers. *Cancer Biol. Ther.* **2011**, *11*, 263–276. [CrossRef]

30. Chang, C.W.; Chen, Y.S.; Chou, S.H.; Han, C.L.; Chen, Y.J.; Yang, C.C.; Huang, C.Y.; Lo, J.F. Distinct subpopulations of head and neck cancer cells with different levels of intracellular reactive oxygen species exhibit diverse stemness, proliferation, and chemosensitivity. *Cancer Res.* **2014**, *74*, 6291–6305. [CrossRef]

31. Gibson, S.B. Chapter Thirteen - Investigating the Role of Reactive Oxygen Species in Regulating Autophagy. *Meth. Enzymol.* **2013**, *528*, 217–235. [CrossRef]

32. Sun, Y.; Zou, H.; Yang, L.; Zhou, M.; Shi, X.; Yang, Y.; Chen, W.; Zhao, Y.; Mo, J.; Lu, Y. Effect on the liver cancer cell invasion ability by studying the associations between autophagy and TRAP1 expression. *Oncol. Lett.* **2018**, *16*, 991–997. [CrossRef] [PubMed]

33. Xie, X.; Zhu, H.; Zhang, J.; Wang, M.; Zhu, L.; Guo, Z.; Shen, W.; Wang, D. Solamargine inhibits the migration and invasion of HepG2 cells by blocking epithelial-to-mesenchymal transition. *Oncol. Lett.* **2017**, *14*, 447–452. [CrossRef] [PubMed]

34. Zheng, J.; Shao, Y.; Jiang, Y.; Chen, F.; Liu, S.; Yu, N.; Zhang, D.; Liu, X.; Zou, L. Tangeretin inhibits hepatocellular carcinoma proliferation and migration by promoting autophagy-related BECLIN1. *Cancer Manag. Res.* **2019**, *11*, 5231–5242. [CrossRef] [PubMed]

35. Furth, E.E.; Sprecher, H.; Fisher, E.A.; Fleishman, H.D.; Laposata, M. An in vitro model for essential fatty acid deficiency: HepG2 cells permanently maintained in lipid-free medium. *J. Lipid Res.* **1992**, *33*, 1719–1726. [CrossRef]

36. Desquiret, V.; Loiseau, D.; Jacques, C.; Douay, O.; Malthièry, Y.; Ritz, P.; Roussel, D. Dinitrophenol-induced mitochondrial uncoupling in vivo triggers respiratory adaptation in HepG2 cells. *Biochim. Biophys. Acta* **2006**, *1757*, 21–30. [CrossRef]

37. Salas-Rojas, M.; Galvez-Romero, G.; Anton-Palma, B.; Acevedo, R.; Blanco-Favela, F.; Aguilar-Setién, A. The coelomic fluid of the sea urchin Tripneustes depressus shows antiviral activity against Suid herpesvirus type 1 (SHV-1) and rabies virus (RV). *Fish Shellfish Immunol.* **2014**, *36*, 158–163. [CrossRef]

38. Soleimani, S.; Mashjoor, S.; Mitra, S.; Yousefzadi, M.; Rezadoost, H. Coelomic fluid of Echinometra mathaei: The new prospects for medicinal antioxidants. *Fish Shellfish Immunol.* **2021**, *117*, 311–319. [CrossRef]

39. Widyananda, M.H.; Pratama, S.K.; Samoedra, R.S.; Sari, F.N.; Kharisma, V.D.; Ansori, A.N.M.; Antonius, Y. Molecular docking study of sea urchin (Arbacia lixula) peptides as multi-target inhibitor for non-small cell lung cancer (NSCLC) associated proteins. *J. Pharm. Pharmacogn. Res.* **2021**, *9*, 484–496.

40. D'Alessio, S.; Buckley, K.M.; Kraev, I.; Hayes, P.; Lange, S. Extracellular Vesicle Signatures and Post-Translational Protein Deimination in Purple Sea Urchin (Strongylocentrotus purpuratus) Coelomic Fluid-Novel Insights into Echinodermata Biology. *Biology* **2021**, *10*, 866. [CrossRef]

41. Ko, K.L.; Mak, L.Y.; Cheung, K.S.; Yuen, M.F. Hepatocellular carcinoma: Recent advances and emerging medical therapies. *F1000Research* **2020**, *9*, 620. [CrossRef]

42. Zimmermann, A. Etiology and Pathogenesis of Hepatocellular Carcinoma: Inflammatory and Toxic Causes. In *Tumors and Tumor-Like Lesions of the Hepatobiliary Tract: General and Surgical Pathology*; Zimmermann, A., Ed.; Springer International Publishing: Cham, Switzerland, 2017; pp. 2931–2959.

43. Wallace, M.C.; Friedman, S.L. Hepatic fibrosis and the microenvironment: Fertile soil for hepatocellular carcinoma development. *Gene Expr.* **2014**, *16*, 77–84. [CrossRef] [PubMed]

44. Alnajjar, A.M.; Elsiesy, H.A. Natural products and hepatocellular carcinoma: A review. *Hepatoma Res.* **2015**, *1*, 119–124. [CrossRef]

45. Jain, D.; Murti, Y.; Khan, W.U.; Hossain, R.; Hossain, M.N.; Agrawal, K.K.; Ashraf, R.A.; Islam, M.T.; Janmeda, P.; Taheri, Y.; et al. Roles of Therapeutic Bioactive Compounds in Hepatocellular Carcinoma. *Oxid. Med. Cell Longev.* **2021**, *2021*, 9068850. [CrossRef]

46. Singh, A.; Verma, S.; Modak, S.B.; Chaturvedi, M.M.; Purohit, J.S. Extra-nuclear histones: Origin, significance and perspectives. *Mol. Cell. Biochem.* **2022**, *477*, 507–524. [CrossRef]

47. Tadayoni Nia, A.; Bazi, Z.; Khosravi, A.; Oladnabi, M. WDR81 Gene Silencing Can Reduce Exosome Levels in Human U87-MG Glioblastoma Cells. *J. Mol. Neurosci.* **2021**, *71*, 1696–1702. [CrossRef]

48. Morita, S.Y.; Deharu, Y.; Takata, E.; Nakano, M.; Handa, T. Cytotoxicity of lipid-free apolipoprotein B. *Biochim. Biophys. Acta* **2008**, *1778*, 2594–2603. [CrossRef] [PubMed]

49. Hall, A.K. Thymosin beta-10 accelerates apoptosis. *Cell. Mol. Biol. Res.* **1995**, *41*, 167–180.

50. Rho, S.B.; Lee, K.W.; Chun, T.; Lee, S.H.; Park, K.; Lee, J.H. The identification of apoptosis-related residues in human thymosin beta-10 by mutational analysis and computational modeling. *J. Biol. Chem.* **2005**, *280*, 34003–34007. [CrossRef]

51. Lee, S.H.; Son, M.J.; Oh, S.H.; Rho, S.B.; Park, K.; Kim, Y.J.; Park, M.S.; Lee, J.H. Thymosin {beta}(10) inhibits angiogenesis and tumor growth by interfering with Ras function. *Cancer Res.* **2005**, *65*, 137–148. [CrossRef]

52. Rodrigo, A.P.; Mendes, V.M.; Manadas, B.; Grosso, A.R.; Alves de Matos, A.P.; Baptista, P.V.; Costa, P.M.; Fernandes, A.R. Specific Antiproliferative Properties of Proteinaceous Toxin Secretions from the Marine Annelid Eulalia sp. onto Ovarian Cancer Cells. *Mar. Drugs* **2021**, *19*, 31. [CrossRef]

53. Lai, J.P.; Yu, C.; Moser, C.D.; Aderca, I.; Han, T.; Garvey, T.D.; Murphy, L.M.; Garrity-Park, M.M.; Shridhar, V.; Adjei, A.A.; et al. SULF1 inhibits tumor growth and potentiates the effects of histone deacetylase inhibitors in hepatocellular carcinoma. *Gastroenterology* **2006**, *130*, 2130–2144. [CrossRef] [PubMed]

54. Frau, M.; Simile, M.M.; Tomasi, M.L.; Demartis, M.I.; Daino, L.; Seddaiu, M.A.; Brozzetti, S.; Feo, C.F.; Massarelli, G.; Solinas, G.; et al. An expression signature of phenotypic resistance to hepatocellular carcinoma identified by cross-species gene expression analysis. *Cell. Oncol.* **2012**, *35*, 163–173. [CrossRef] [PubMed]

55. Sheng, S.; Jiwen, W.; Dexiang, Z.; Bohao, Z.; Yueqi, W.; Han, L.; Xiaoling, N.; Tao, S.; Liu, H. DMBT1 suppresses progression of gallbladder carcinoma through PI3K/AKT signaling pathway by targeting PTEN. *Biosci. Biotechnol. Biochem.* **2019**, *83*, 2257–2264. [CrossRef] [PubMed]

56. Sun, G.; Peng, B.; Xie, Q.; Ruan, J.; Liang, X. Upregulation of ZBTB7A exhibits a tumor suppressive role in gastric cancer cells. *Mol. Med. Rep.* **2018**, *17*, 2635–2641. [CrossRef] [PubMed]

57. Xiang, T.; Tang, J.; Li, L.; Peng, W.; Du, Z.; Wang, X.; Li, Q.; Xu, H.; Xiong, L.; Xu, C.; et al. Tumor suppressive BTB/POZ zinc-finger protein ZBTB28 inhibits oncogenic BCL6/ZBTB27 signaling to maintain p53 transcription in multiple carcinogenesis. *Theranostics* **2019**, *9*, 8182–8195. [CrossRef] [PubMed]

58. He, J.; Wu, M.; Xiong, L.; Gong, Y.; Yu, R.; Peng, W.; Li, L.; Li, L.; Tian, S.; Wang, Y.; et al. BTB/POZ zinc finger protein ZBTB16 inhibits breast cancer proliferation and metastasis through upregulating ZBTB28 and antagonizing BCL6/ZBTB27. *Clin. Epigenetics* **2020**, *12*, 82. [CrossRef]

59. Hadari, Y.R.; Arbel-Goren, R.; Levy, Y.; Amsterdam, A.; Alon, R.; Zakut, R.; Zick, Y. Galectin-8 binding to integrins inhibits cell adhesion and induces apoptosis. *J. Cell Sci.* **2000**, *113*, 2385–2397. [CrossRef]

60. Autelli, R.; Ullio, C.; Prigione, E.; Crepaldi, S.; Schiavone, N.; Brunk, U.T.; Capaccioli, S.; Baccino, F.M.; Bonelli, G. Divergent pathways for TNF and C(2)-ceramide toxicity in HTC hepatoma cells. *Biochim. Biophys. Acta* **2009**, *1793*, 1182–1190. [CrossRef]

61. Peng, X.; Wen, Y.; Zha, L.; Zhuang, J.; Lin, L.; Li, X.; Chen, Y.; Liu, Z.; Zhu, S.; Liang, J.; et al. TRIM45 Suppresses the Development of Non-small Cell Lung Cancer. *Curr. Mol. Med.* **2020**, *20*, 299–306. [CrossRef]

62. Zhang, J.; Zhang, C.; Cui, J.; Ou, J.; Han, J.; Qin, Y.; Zhi, F.; Wang, R.F. TRIM45 functions as a tumor suppressor in the brain via its E3 ligase activity by stabilizing p53 through K63-linked ubiquitination. *Cell Death Dis.* **2017**, *8*, e2831. [CrossRef]

63. Wallner, C.; Drysch, M.; Becerikli, M.; Jaurich, H.; Wagner, J.M.; Dittfeld, S.; Nagler, J.; Harati, K.; Dadras, M.; Philippou, S.; et al. Interaction with the GDF8/11 pathway reveals treatment options for adenocarcinoma of the breast. *Breast* **2018**, *37*, 134–141. [CrossRef]

64. Drysch, M.; Schmidt, S.V.; Becerikli, M.; Reinkemeier, F.; Dittfeld, S.; Wagner, J.M.; Dadras, M.; Sogorski, A.; von Glinski, M.; Lehnhardt, M.; et al. Myostatin Deficiency Protects C2C12 Cells from Oxidative Stress by Inhibiting Intrinsic Activation of Apoptosis. *Cells* **2021**, *10*, 1680. [CrossRef] [PubMed]

65. Tanaka, M.; Shimbo, T.; Kikuchi, Y.; Matsuda, M.; Kaneda, Y. Sterile alpha motif containing domain 9 is involved in death signaling of malignant glioma treated with inactivated Sendai virus particle (HVJ-E) or type I interferon. *Int. J. Cancer* **2010**, *126*, 1982–1991. [CrossRef] [PubMed]

66. Kuroda, Y.; Aishima, S.; Taketomi, A.; Nishihara, Y.; Iguchi, T.; Taguchi, K.; Maehara, Y.; Tsuneyoshi, M. 14-3-3sigma negatively regulates the cell cycle, and its down-regulation is associated with poor outcome in intrahepatic cholangiocarcinoma. *Hum. Pathol.* **2007**, *38*, 1014–1022. [CrossRef] [PubMed]

67. Chen, C.; Tian, F.; Lu, L.; Wang, Y.; Xiao, Z.; Yu, C.; Yu, X. Characterization of Cep85 - a new antagonist of Nek2A that is involved in the regulation of centrosome disjunction. *J. Cell Sci.* **2015**, *128*, 3290–3303. [CrossRef] [PubMed]

68. Yu, Y.; Li, Y.; Zhou, L.; Yang, G.; Wang, M.; Hong, Y. Cryptochrome 2 (CRY2) Suppresses Proliferation and Migration and Regulates Clock Gene Network in Osteosarcoma Cells. *Med. Sci. Monit.* **2018**, *24*, 3856–3862. [CrossRef]

69. Sawada, G.; Ueo, H.; Matsumura, T.; Uchi, R.; Ishibashi, M.; Mima, K.; Kurashige, J.; Takahashi, Y.; Akiyoshi, S.; Sudo, T.; et al. CHD8 is an independent prognostic indicator that regulates Wnt/β-catenin signaling and the cell cycle in gastric cancer. *Oncol. Rep.* **2013**, *30*, 1137–1142. [CrossRef]

70. Su, J.; You, P.; Zhao, J.P.; Zhang, S.L.; Song, S.H.; Fu, Z.R.; Ye, L.W.; Zi, X.Y.; Xie, D.F.; Zhu, M.H.; et al. A potential role for the homeoprotein Hhex in hepatocellular carcinoma progression. *Med. Oncol.* **2012**, *29*, 1059–1067. [CrossRef]

71. Marfil, V.; Blazquez, M.; Serrano, F.; Castell, J.V.; Bort, R. Growth-promoting and tumourigenic activity of c-Myc is suppressed by Hhex. *Oncogene* **2015**, *34*, 3011–3022. [CrossRef]

72. Ni, S.; Hu, J.; Duan, Y.; Shi, S.; Li, R.; Wu, H.; Qu, Y.; Li, Y. Down expression of LRP1B promotes cell migration via RhoA/Cdc42 pathway and actin cytoskeleton remodeling in renal cell cancer. *Cancer Sci.* **2013**, *104*, 817–825. [CrossRef]

73. Beer, A.G.; Zenzmaier, C.; Schreinlechner, M.; Haas, J.; Dietrich, M.F.; Herz, J.; Marschang, P. Expression of a recombinant full-length LRP1B receptor in human non-small cell lung cancer cells confirms the postulated growth-suppressing function of this large LDL receptor family member. *Oncotarget* **2016**, *7*, 68721–68733. [CrossRef] [PubMed]

74. Zhong, X.; Liu, D.; Hao, Y.; Li, C.; Hao, J.; Lin, C.; Shi, S.; Wang, D. The expression of TET3 regulated cell proliferation in HepG2 cells. *Gene* **2019**, *698*, 113–119. [CrossRef] [PubMed]

75. Fu, D.; Jordan, J.J.; Samson, L.D. Human ALKBH7 is required for alkylation and oxidation-induced programmed necrosis. *Genes Dev.* **2013**, *27*, 1089–1100. [CrossRef] [PubMed]

76. Na, J.; Jung, J.; Bang, J.; Lu, Q.; Carlson, B.A.; Guo, X.; Gladyshev, V.N.; Kim, J.; Hatfield, D.L.; Lee, B.J. Selenophosphate synthetase 1 and its role in redox homeostasis, defense and proliferation. *Free Radic. Biol. Med.* **2018**, *127*, 190–197. [CrossRef] [PubMed]

77. Zhang, L.; Rong, W.; Ma, J.; Li, H.; Tang, X.; Xu, S.; Wang, L.; Wan, L.; Zhu, Q.; Jiang, B.; et al. Comprehensive Analysis of DNA 5-Methylcytosine and N6-Adenine Methylation by Nanopore Sequencing in Hepatocellular Carcinoma. *Front. Cell. Dev. Biol.* **2022**, *10*, 827391. [CrossRef]

78. Zhang, Y.; Tong, G.H.; Wei, X.X.; Chen, H.Y.; Liang, T.; Tang, H.P.; Wu, C.A.; Wen, G.M.; Yang, W.K.; Liang, L.; et al. Identification of Five Cytotoxicity-Related Genes Involved in the Progression of Triple-Negative Breast Cancer. *Front. Genet.* **2022**, *12*, 723477. [CrossRef]

79. Manshouri, R.; Coyaud, E.; Kundu, S.T.; Peng, D.H.; Stratton, S.A.; Alton, K.; Bajaj, R.; Fradette, J.J.; Minelli, R.; Peoples, M.D.; et al. ZEB1/NuRD complex suppresses TBC1D2b to stimulate E-cadherin internalization and promote metastasis in lung cancer. *Nat. Commun.* **2019**, *10*, 5125. [CrossRef]

80. Fatahala, S.S.; Sayed, A.I.; Mahgoub, S.; Taha, H.; El-Sayed, M.-I.k.; El-Shehry, M.F.; Awad, S.M.; Abd El-Hameed, R.H. Synthesis of Novel 2-Thiouracil-5-Sulfonamide Derivatives as Potent Inducers of Cell Cycle Arrest and CDK2A Inhibition Supported by Molecular Docking. *Int. J. Mol. Sci.* **2021**, *22*, 11957. [CrossRef]

81. Han, Z.; Zhao, X.; Zhang, E.; Ma, J.; Zhang, H.; Li, J.; Xie, W.; Li, X. Resistomycin Induced Apoptosis and Cycle Arrest in Human Hepatocellular Carcinoma Cells by Activating p38 MAPK Pathway In Vitro and In Vivo. *Pharmaceuticals* **2021**, *14*, 958. [CrossRef]

82. Ren, Y.; Ruan, Y.; Cheng, B.; Li, L.; Liu, J.; Fang, Y.; Chen, J. Design, synthesis and biological evaluation of novel acridine and quinoline derivatives as tubulin polymerization inhibitors with anticancer activities. *Bioorg. Med. Chem.* **2021**, *46*, 116376. [CrossRef]

83. Zhu, C.; Zhao, M.; Fan, L.; Cao, X.; Xia, Q.; Zhou, J.; Yin, H.; Zhao, L. Chitopentaose inhibits hepatocellular carcinoma by inducing mitochondrial mediated apoptosis and suppressing protective autophagy. *Bioresour. Bioprocess.* **2021**, *8*, 4. [CrossRef]

84. Jiang, J.; Briedé, J.J.; Jennen, D.G.; Van Summeren, A.; Saritas-Brauers, K.; Schaart, G.; Kleinjans, J.C.; de Kok, T.M. Increased mitochondrial ROS formation by acetaminophen in human hepatic cells is associated with gene expression changes suggesting disruption of the mitochondrial electron transport chain. *Toxicol. Lett.* **2015**, *234*, 139–150. [CrossRef] [PubMed]

85. Liu, B.; Tan, X.; Liang, J.; Wu, S.; Liu, J.; Zhang, Q.; Zhu, R. A reduction in reactive oxygen species contributes to dihydromyricetin-induced apoptosis in human hepatocellular carcinoma cells. *Sci. Rep.* **2014**, *4*, 7041. [CrossRef] [PubMed]

Journal of
Marine Science and Engineering

Microscopic Anatomy of the Lining of Hemal Spaces in the Penaeid Shrimp, *Sicyonia ingentis*

Rachel Brittany Sidebottom, Sabi Bang and Gary Martin *

Department of Biology, Occidental College, Los Angeles, CA 90041, USA; rsidebottom@oxy.edu (R.B.S.); sbang2@oxy.edu (S.B.)
* Correspondence: gmartin@oxy.edu; Tel.: +1-323-259-2890

Abstract: The purpose of this paper is to present a morphological description of three different types of acellular material lining hemal spaces in a shrimp, providing a background for addressing future questions. The vasculature of the penaeid shrimp, *Sicyonia ingentis*, includes vessels leading from the heart into arteries which branch and expand into sinuses before returning hemolymph back to the heart. Early work showed that an endothelium was absent, and a basement membrane (BM) separated tissues from the hemolymph. Therefore, it was suggested that hemocytes could identify anything other than the BM as a "foreign" entity. This study demonstrates three major types of acellular material lining the hemal spaces of *S. ingentis*. Cardiomyocytes, digestive gland tubules, and abdominal muscle fibers are covered by BMs. Major arteries are lined by a fibrillin-like fibrous material. Finally, sheaths of collagenous connective tissues cover the heart and digestive gland as well as the outer surface of arteries, the gut, and gonad. Our understanding of hemocyte receptors and extracellular matrices in general have greatly expanded but the biochemical composition of the matrices lining crustacean hemal spaces, their role in regulating nutrient uptake, and the cells responsible for their deposition deserve further attention.

Keywords: crustacean; vasculature; extracellular matrix; basement membrane; connective tissue sheath; fibrillin

Citation: Sidebottom, R.B.; Bang, S.; Martin, G. Microscopic Anatomy of the Lining of Hemal Spaces in the Penaeid Shrimp, *Sicyonia ingentis*. *J. Mar. Sci. Eng.* **2021**, *9*, 862. https://doi.org/10.3390/jmse9080862

Academic Editor: Qi Li

Received: 8 June 2021
Accepted: 9 August 2021
Published: 11 August 2021

Publisher's Note: MDPI stays neutral with regard to jurisdictional claims in published maps and institutional affiliations.

1. Introduction

Decapod crustaceans have a "semi-open" [1] or "incompletely closed" [2] circulatory system, meaning that hemolymph is not completely contained within vessels as it travels throughout the body. Although arteries extend from the heart and branch into smaller tubes, the vessels eventually end and expand into sinuses before the hemolymph drains back into the heart; veins are absent [1,3]. Shrimp hemal spaces lack an endothelium and instead are lined by an extracellular matrix [4]. Circulating fluid is called hemolymph because there is no separation of blood from lymph, and it bathes the surfaces of tissues and organs directly [2].

In 1970, it was suggested that circulating hemocytes (blood cells) might identify non-self materials as anything differing from the matrix lining hemal spaces, which was called a basement membrane (BM) [5]. In 1984, the idea was expanded to suggest that hemocytes moving past these surfaces receive signals regarding the presence of foreign materials [6]. Our understanding of the repertoire of receptors on hemocytes has greatly expanded with continued research on invertebrate innate immunity and now includes toll-like receptors [7,8], integrins [9–11], lectins [12], and pattern-recognition motifs [13].

Our understanding of BMs and extracellular matrices has also greatly expanded mostly in vertebrate systems, while knowledge on the extracellular matrices lining hemal spaces in crustaceans has received little attention. Typically, the morphology of these primarily acellular layers is only noted in passing in descriptions focusing on the cellular components of particular tissues [14–17], only a single type of matrix has been described in each study, and a variety of terms have been used such as BMs, basal laminae, internal/external laminae, and tunica intimae.

This paper will demonstrate that three types of acellular materials line hemal spaces in different regions of the penaeid shrimp *Sicyonia ingentis*: BMs, elastic fibrillin-like layers, and connective tissue sheaths. The first type of acellular material, basement membranes, has been well studied in vertebrates where they provide mechanical support to epithelial cells [18–20] while influencing macromolecule exchange between blood and tissues [19]. There are considerable morphological variations seen in invertebrates [21] and the use of the term basement membrane was encouraged "as a general descriptive and comprehensive word to be used in light microscopy, biochemistry and electron microscopy". At the light microscope level, BMs appear as thin layers separating epithelia from underlying connective tissue and stain positive with PAS. At the TEM level, three layers may be seen; an electron-lucent lamina rara beneath the plasma membrane followed by the lamina densa and a second electron-lucent zone called the lamina reticularis, which blends into the underlying connective tissue when present [21]. The major component of BMs is collagen type IV, localized within the lamina densa [18,21–24].

An important difference between BMs in vertebrate vessels and the ones lining crustacean (as well as most other invertebrates) hemal spaces is that in the latter the BM faces an epithelium on one side and a hemal space on the other. In vertebrate vessels, a BMs lies between the basal surface of the endothelium and the underlying connective or muscular tissues. In both vertebrates and invertebrates, BMs also surround specialized cells such as adipose cells and striated and cardiac muscle cells [25–31].

A second type of matrix is restricted to the inner lining of large arteries extending from the heart in shrimp [17], lobsters [32], and crab [33]. It has been called the tunica interna (internal lamina) [4,17], it is thicker than a BM, and is composed of a weave of microfibrils. In the shrimp, it stains with histological dyes for elastin [17], but not so in the crab [33]. Work on lobsters has demonstrated it reacts with antibodies to mammalian fibrillin [32,34,35], a glycoprotein involved with the deposition of tropoelastin during the formation of elastic fibers [36–38]. The fibrils themselves have elastic properties in the lobster artery and assist in propagation of blood following contraction of the heart [32].

The third acellular matrix lining hemal spaces is a connective tissue sheath which has been described as "fundamentally an accumulation of collagen fibrils" [39]. In that study, the term tunica externa was used, and it was noted that the layers of collagen are oriented primarily longitudinally to blood flow with sparse, interspersed fibroblast-like cells. The fibrils surround the outer surfaces of blood vessels and the gut (fore-, mid-, and hind-gut), as well as forming sheaths surrounding the entire collection of digestive gland tubules [40], and nerves [41–44].

The goal of this study is to make these linings the focus of a morphological description presenting the diversity of these matrices past which hemocytes and hemolymph circulate within a single crustacean, the penaeid shrimp *Sicyonia ingentis*. This will set the stage for future work to characterize the biochemical nature of these layers and their roles in nutrient exchange.

2. Materials and Methods

2.1. Collection and Maintenance of Shrimp

Adult, sexually mature ridgeback prawns, *Sicyonia ingentis*, were collected in 100 m of water off the coast of Palos Verdes, California by otter trawls in April 2019. Shrimp averaged 14.5 ± 3.3 g and were in molt stage D [45]. The shrimp were kept in aquaria with salt-water at 15 °C and 33 ppt salinity.

2.2. Tissue Collection and Processing

The stomach, digestive gland (outer covering of entire organ as well as individual tubules), midgut trunk, gonad, heart, dorsal abdominal artery, and muscle from the dorsal wall of the 4th abdominal segment were carefully removed from the body of the shrimp and immersed in fixatives (see below).

For light microscopy (LM), tissues and organs were cut into pieces 1 cm^3 and fixed for a minimum of 3 h in half-strength Karnovsky's fixative (2% formaldehyde, 2.5% glutaraldehyde in 0.1 M sodium cacodylate pH 7.4) and processed for standard paraffin embedding. Thick sections (8 μm) were stained with hematoxylin and eosin, periodic acid-Schiff procedure (Sigma-Aldrich, St. Louis, MO, USA; kit 395B), and Masson's trichrome stain (Sigma-Aldrich St. Louis, MO, USA; kit HT15), using standard procedures [46].

For scanning electron microscopy (SEM), similar sized pieces of tissues and organs were cut adjacent to those selected for LM and fixed in 3% glutaraldehyde in 0.1 M sodium cacodylate pH 7.4 containing 12% glucose for 3 h. Samples were then dehydrated in a graded series of ethanol, dried using hexamethyldisilane (Pella 18605), coated with gold in a Technics Hummer 2, and examined in a Phenom SEM.

For transmission electron microscopy (TEM), tissues were fixed overnight in the same fixative used for SEM, washed 10 min in buffer (0.1 M sodium cacodylate with 24% sucrose), post-fixed 1 h in 1% OsO_4 in 0.1 M sodium cacodylate buffer, stained *en bloc* for 1 h in 3% uranyl acetate in 0.1 M sodium acetate buffer pH 4.5, dehydrated through a series of ethanol, and infiltrated and embedded in Spurr's plastic [47]. Semi-thin sections (0.5 μm) were stained with methylene blue and thin sections (70 nm) were stained with lead citrate and viewed in a Zeiss Laborlux 12 light microscope and a Zeiss EM 109 transmission electron microscope (TEM), respectively.

2.3. Measurements

Ten shrimp were examined in this study; five were processed for paraffin embedding and examination by LM, five were processed for TEM and SEM. For LM and TEM, one block from each sample was examined to produce ten images used for measurements. For SEM, three measurements were made on a sample from each tissue. Measurements reported in the text are presented as the range, and the mean ± one standard deviation.

3. Results

3.1. Distribution of Hemal Spaces

Blood is pumped from the heart into several large arteries that branch into narrower vessels, which eventually end such that the hemolymph flows into spaces around and directly bathes tissues and organs (Figure 1). The linings of the major vessels and the external surfaces of the tissues and organs are described below.

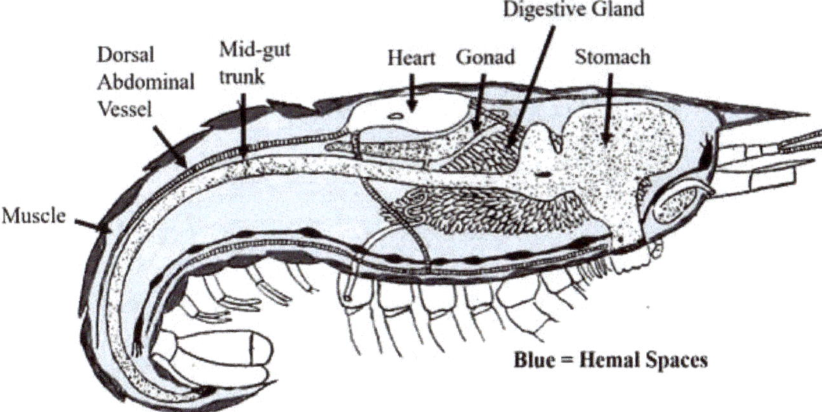

Figure 1. Schematic of shrimp anatomy displaying organs adjacent to hemal spaces. Blood also fills the heart and dorsal abdominal artery and surrounds individual tubules within the digestive gland.

3.2. Fibrillin-like Lining of Major Arteries

The wall of the dorsal abdominal artery measures between 50.2–65.5, ($\bar{x}$ = 58.2 ± 0.8) μm thick and is composed of the following layers (Figure 2A); an innermost fibrillin-like layer, a thin layer of muscle (6.4–9.7, ($\bar{x}$ = 7.3 ± 0.6 μm), two additional fibrillin layers (4.0–10.4 ± 0.5 μm) separated by loose connective tissue (10.4–12.2, ($\bar{x}$ = 11.6 ± 0.8 μm) and an outer bounding connective tissue sheath (see Section 3.4). The inner fibrillin lining (Figure 2A) is thin (1.0–1.5, ($\bar{x}$ = 1.2 ± 0.3 μm) and stains with eosin and PAS but does not stain with trichrome for collagen. When viewed by SEM (Figure 2B,C) the surface is smooth with folds ranging in width between 16.6 μm and 23.2 μm, which may be due to shrinkage or contraction during processing. At higher magnification (Figure 2C), the smooth surface presents low profile ridges (0.4 μm in width) running primarily parallel to the long axis of the vessel. The ridges have the same diameter as pseudopodal extensions from attached hemocytes and may represent remnants of detached cells or exposed filaments that make up the inner layer of the vessel wall. When examined using TEM, the inner fibrillin layer is composed of fibrils (22.2–29.8, $\bar{x}$ = 26.4 ± 1.4 nm diameter) forming a meshwork (Figure 2D).

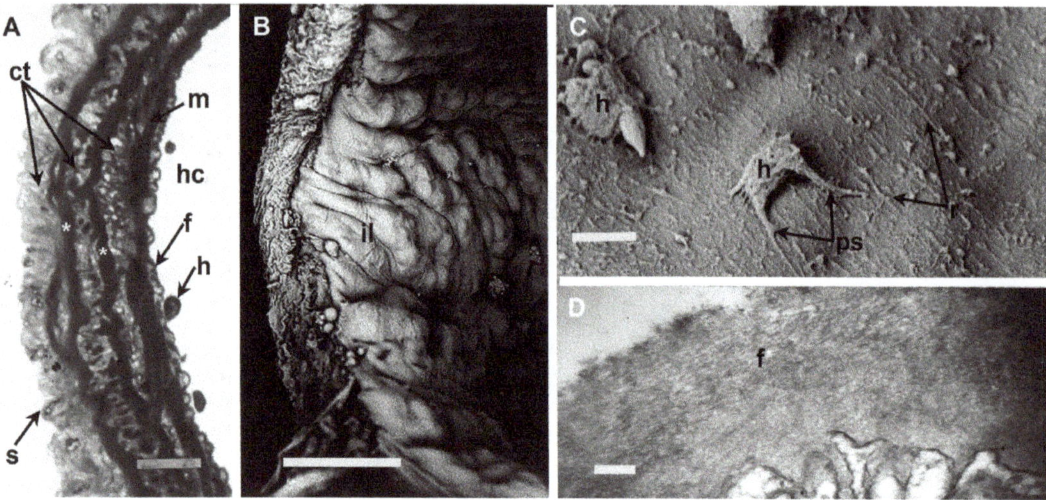

Figure 2. Inner lining of the dorsal abdominal artery. (**A**) Light micrograph of cross section through vessel wall showing inner lining of fibrillin-like material (f) separating the hemocoel (hc) and a hemocyte (h) from three wall layers of connective tissue (ct), separated by two additional layers of fibrillin (*), each outside of a muscle layer (m). An outer connective tissue sheath (s) surrounds the entire vessel. (**B**) Low magnification view of inner vessel lining (il) showing folded surface. (**C**) Higher magnification view of inner fibrillin surface and attached hemocytes (h) with extended pseudopodia (ps) tapering to diameters like small ridges (r) running parallel to long axis of vessel. (**D**) Transmission electron micrograph showing the weave of filaments in the fibrillin layer (f) covering a ridge on the surface. Scale bar (**A**) 50 μm, (**B**) 50 μm, (**C**) 10 μm, (**D**) 0.25 μm.

3.3. Tissues Lined by BM

Individual heart myocytes, tubules of the digestive gland, and abdominal muscle fibers are separated from hemal spaces by BMs too thin to be resolved clearly at the LM level. All stain positive with PAS and faintly with eosin and trichrome. SEM of cardiac myocytes shows a surface with longitudinal folds as well as periodic constrictions every 4.5–9.0 μm, the latter possibly due to contraction during processing (Figure 3A). At higher magnification, the BM seen between constrictions is smooth with widely spaced and randomly oriented filaments (0.3–0.7, $\bar{x} = 0.5 \pm 0.2$ μm diameter). Hemocytes were commonly seen bound to the surface either singularly or in clusters of two to four hemocytes (7.2–8.5 μm diameter; Figure 3B).

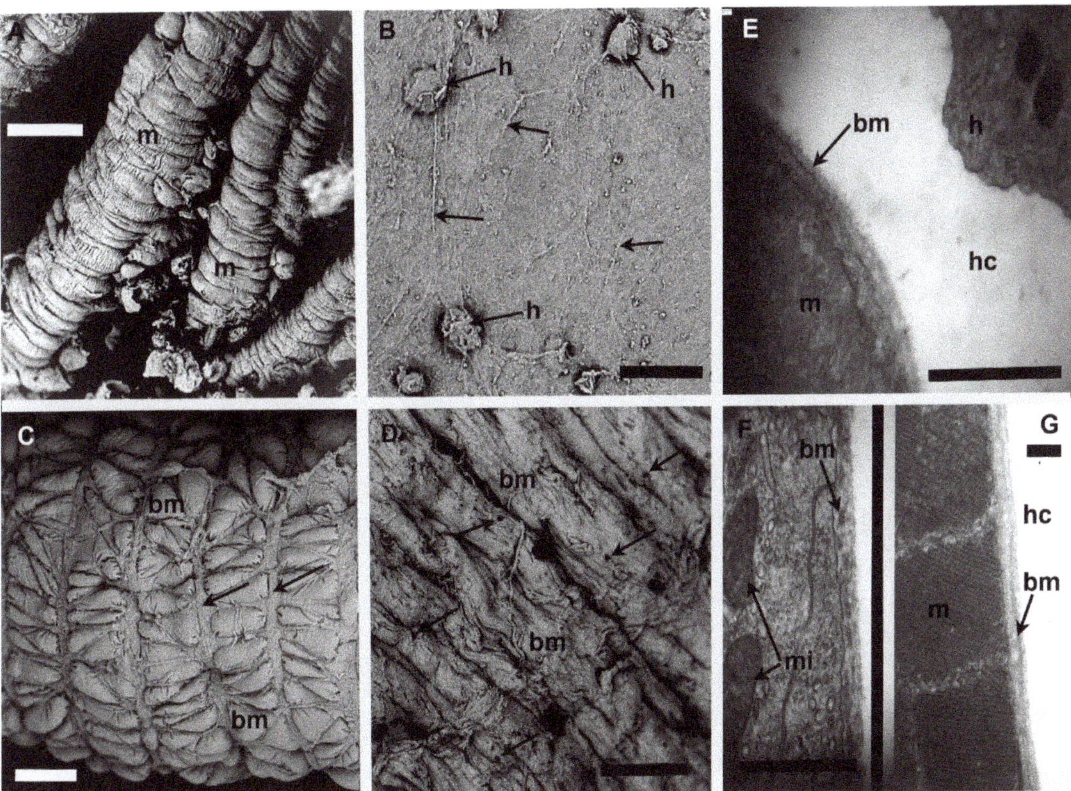

Figure 3. Tissues encased within basement membranes. (**A**) Scanning electron micrograph (SEM) of individual contracted cardiac myocytes (m). (**B**) Higher magnification SEM showing BM surrounding myocytes (which is the same as seen covering individual digestive gland tubules and abdominal muscle fibers) with attached hemocytes (h) and fine ridges (arrows). (**C**) Low magnification SEM of surface of individual tubule from digestive gland with basement membrane (bm) constricted periodically by thicker circumferentially arranged contractile cells (arrows) and thin cells running longitudinally causing the BM to form bulges. (**D**) SEM showing surface of adjacent fibers from abdominal muscle covered by a basement membrane (bm). An occasional collagen fiber and small pores (arrows) are seen on the surface. (**E**) Transmission electron micrograph (TEM) showing thin basement membrane (bm) covering a myocyte (m) and the edge of hemocyte (h) in adjacent hemocoel (hc). (**F**) TEM showing lamina densa layers of basement membrane (bm) covering the basal region of digestive gland cells with mitochondria (m) and vesicles. (**G**) TEM of basement membrane (bm) separating hemocoel (hc) and abdominal muscle cell (mc). Scale bar (**A**) 30 μm, (**B**) 15 μm, (**C**) 25 μm, (**D**) 100 μm, (**E**) 1 μm, (**F**) 0.5 μm, (**G**) 0.5 μm.

Although the entire digestive gland is surrounded by a connective tissue sheath (see Section 3.4), each cylindrical tubule of the digestive gland is encased within a BM (Figure 3C). Approximately every 16.7 μm there is a band of contractile cells (1.1–1.4, $\bar{x} = 1.2 \pm 0.2$ μm wide) oriented around the circumference of each tubule. Additional contractile cells oriented longitudinally subdivide each cylinder into approximately 50–60 ovoid or triangular bulges (14.3–16.7, $\bar{x} = 15.0 \pm 2.1 \times 5.1$–7.5, $\bar{x} = 6.2 \pm 1.1$ μm). Between the constrictions, the BM is smooth and appears the same as that shown in Figure 3B. Small aggregations of two to five hemocytes are commonly seen attached to the surface especially in the depressions between adjacent bulges and the contractile cells.

Individual abdominal muscle cells appear in SEM as elongate cylinders approximately 50–90 µm in diameter with a wrinkled BM coating their surfaces (Figure 3D). Randomly spaced pores measuring 5.3–10.4, ($\bar{x} = 7.8 \pm 1.4$) µm diameter are common, as are individual and small clusters of attached hemocytes.

When examined by TEM, the BMs of these three tissues are similar (Figure 3E–G). The lamina densa in the BM surrounding each cardiac myocyte is a single fuzzy, moderately electron-dense layer composed of fibers ranging between 30–40, and averaging 33 ± 2 nm in diameter (Figure 3E). The lamina densa in the BM around each digestive gland tubule forms an incomplete layer of electron-dense bodies (Figure 3F). In the BM of abdominal muscles, the lamina densa consists of two electron-dense layers, one thicker (20–40, $\bar{x} = 33 \pm 2$ nm) and one thinner layer (approximately 40 nm; Figure 3G). All BMs lie directly against the plasma membrane.

3.4. Tissues Lined by Connective Tissue Sheaths

Although individual cardiac myocytes, tubules of the digestive gland, and skeletal muscle cells are encased within BMs, the collection of individual cells and tubules is surrounded by a connective tissue sheath that encases each organ and separates it from a hemal space. The term sheath is used when multiple collagenous layers invest an organ. Similar sheaths were found covering the surface of the dorsal abdominal artery, the foregut, the midgut trunk, and the gonad. All sheaths display a rough, irregular surface facing the hemolymph, clearly shown using SEM. Figure 4A shows the irregular surface of the sheath enclosing the heart composed of strands of collagenous fibers (typically 4.4–7.5, $\bar{x} = 5.2 \pm 1.2$ µm diameter) forming a meshwork weaving around pores and depressions that both range in diameter between 4.5 µm and 22.7 µm. Pores appear as empty channels leading deeper into the sheath whereas depressions may have a bottom shelf. Hemocytes (averaging 6.7 ± 2.1 µm in diameter) are frequently observed on the surface of the outer investing layer of the heart, occasionally in clusters up to 30 cells.

The surface of the sheath covering the collection of digestive gland tubules is wrinkled with a relatively smooth surface between folds (Figure 4B). Pores were small (1.2–4.7, $\bar{x} = 2.8 \pm 0.5$ µm diameter) and rare. Hemocytes were observed on the surface of the sheath and bound to pores. A fortuitous tear in the sheath allows visualization of the constricted BM coating an individual digestive gland tubule lying beneath the sheath (Figure 4B).

The outer surface of the dorsal abdominal artery and midgut trunk are also covered by connective tissue sheaths, which show the greatest degree of surface irregularities (Figure 4C,D); the sheaths are highly folded. The surface of the dorsal abdominal artery has ridges oriented primarily parallel to the long axis of the vessel (Figure 4C). A small vessel is shown branching from the dorsal artery and has the same appearance. The folds on the surface of the midgut trunk appear as overlapping flaps with a width and height ranging between 4–8 and 5–9 µm, respectively, with a thickened ridge over the apex and pores, larger than seen in other tissue, up to 35 µm in diameter are present (Figure 4D). Fewer hemocytes are seen on the surface of the midgut trunk as compared to the dorsal abdominal artery, but they are always present.

When observed by SEM, the sheath surrounding the gonad lacks the high-profile ridges seen on the dorsal abdominal artery and midgut trunk; the surface shows multiple folds running both along the longitudinal and circumferential axes (Figure 4E). Pores were not observed and individual hemocytes were occasionally present. The sheath covering the foregut (Figure 4F) is morphologically similar to the sheath over the gonad; it is flat with a roughened surface penetrated by numerous small, randomly distributed pores (7.9–23.6 µm in diameter). Hemocytes are common on this surface and are found within many of the holes.

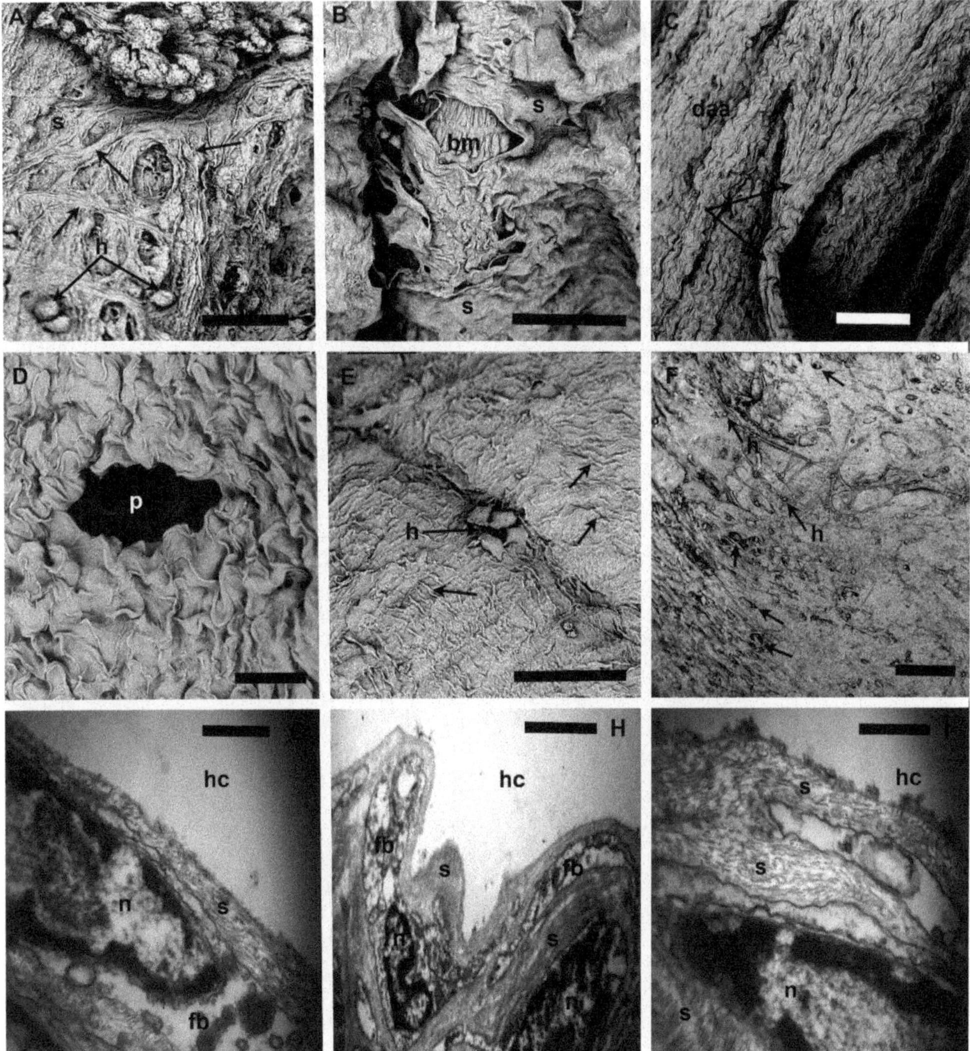

Figure 4. Tissues covered by a connective tissue sheath. (**A**) SEM of sheath (s) overlying a collection of myocytes composing the heart with tracts of collagen fibers (arrows) forming a weave often around ovoid depressions opening to lower layers of the sheath. Hemocytes (h) are common either individually or in clusters. (**B**) SEM of connective tissue sheath (s) covering the digestive gland with an artifactual tear providing a look through the sheath at the underlying basement membrane (bm) covering an individual tubule. (**C**) SEM showing wrinkled surface of outer sheath of connective tissue covering the dorsal abdominal artery (daa) with a smaller vessel (arrows) branching off to one side. (**D**) SEM showing wrinkled surface of connective tissue sheath covering the midgut trunk with an ovoid pore (p) extending into deeper layers of the gut wall. (**E**) SEM of acellular sheath of collagenous material covering the outer surface of the gonad with low profile ridges (arrows) and a small cluster of hemocytes (h) in a fold. (**F**) SEM of surface of foregut covered by a sheath of collagenous material perforated by numerous pores (arrows) opening to deeper layers of the sheath. Bound hemocytes (h) are common. (**G**) TEM showing collagenous layers (s) separated by a fibroblast-like cell (fb). n, nucleus. (**H**) TEM showing folder layers of collagenous sheath (s) separating the hemocoel (hc) from a fibroblast-like cell (fb) in the outer layer of the midgut trunk. n, nucleus. (**I**) TEM showing layers of connective tissue sheaths (s) separating hemocoel (hc) from the nuclear (n) region of fibroblast-like cell. Scale bar (**A**) 25 µm, (**B**) 40 µm, (**C**) 60 µm, (**D**) 15 µm, (**E**) 25 µm, (**F**) 30 um, (**G**) 20 µm, (**H**) 25 um, (**I**) 0.5 µm.

Using TEM, the thickness of the investing sheaths is difficult to measure because collagenous layers lining the hemal spaces appear identical to those that merge with the connective tissue of the walls (Figure 4G–I). An intervening, incomplete layer of fibroblast-like cells may or may not be present. Measuring only the collagenous layers between the cells and the hemal spaces suggests the sheath around the gonad was the thinnest (1.5 ± 1.2 μm) and that investing the foregut the thickest (3.1 ± 1.4 μm). All sheaths examined by TEM showed the same morphology; bundles of collagen fibers separated by electron-lucent spaces with a small number of elongate fibroblast-like cells forming incomplete layers. The extensions of these cells may be extremely thin (0.16 μm) and often appear empty. The collagen fibers form a weave over the surface of each organ. The diameter of individual fibers ranges from 15.9–21.8, $\bar{x}$ = 20.2 ± 1.1 nm in diameter and display a periodicity of approximately 62 ± 2.4 nm. The nucleus of the fibroblast-like cells has patches of heterochromatin along the inner border of the nuclear membrane, sparse RER is present, and Golgi bodies are rare.

4. Discussion

In this study, we describe three distinct types of acellular matrices lining the hemal spaces of the shrimp *Sicyonia ingentis*. First, the dorsal abdominal artery is lined with a fibrillin-like interna. Second, tissues where an epithelium (such as digestive gland tubules) and individually hemal-bathed cells (such as cardiac myocytes and muscle cells) are separated from the hemal spaces by a thin BM. Third, collections of muscle cells and digestive gland tubules are separated from the hemolymph by sheaths of connective tissue which are also present on the outer surfaces of the arteries, gonad, and gut.

Major crustacean arteries are lined with a fibrillar, non-collagenous material as observed in the ophthalmic artery [17], the dorsal abdominal artery of the shrimp (present study), and in the aorta of the lobster *Homarus americanus* [34]. The lobster system has been further studied biochemically to characterize the fibrils as the glycoprotein fibrillin previously known from skin and connective tissues in vertebrates [32]. Its elasticity was demonstrated using stress-strain models [32]. Its presence in the arteries may augment hemolymph propulsion initiated by contractions of the heart.

Once hemolymph flows beyond the large arteries into smaller vessels, there is a transition from a lining of fibrillin to a BM [39]. BMs were also found separating individual cardiac myocytes and abdominal muscle fibers from hemal spaces. In the heart, individual cardiac myocytes are encased within BMs and the entire collection of contractile cardiac cells is surrounded by a connective tissue sheath. Likewise, tubules within that digestive gland of the shrimp rest on a BM that is directly bathed by hemolymph, in agreement with a report on the digestive gland in the lobster [40]. In that study, the sheet of connective tissue encasing the collection of tubules was also described as a "typical, fibrous, connective tissue, containing hemal sinuses and a variety of cell types embedded in a collagenous matrix" [40] as we have described the connective tissue sheath in this paper. The gonad in the shrimp and lobster is also composed of epithelial cells separated from the hemal space by an acellular layer, which was originally referred to as a thick BM [48]. However, we found a thicker layer in the shrimp composed of multiple layers of banded collagen fibers and prefer to characterize it as a connective tissue sheath much like the sheath surrounding the dorsal abdominal artery and midgut trunk. Other organs utilizing an investing layer of BM but not examined in this study include the vessels of the Oka organ [14,17], hematopoietic tissues in shrimp [15], and vessels within the compound eyes of a variety of crustaceans [49]. A fourth organ using BMs is the gill where a study on the shrimp *Caridina japonica* described the gill lamellae being bounded apically by the exoskeleton and basally by a basal lamina lining the vascular lumen [50]. The BMs are the thinnest acellular layers lining hemal spaces, which may allow for the most rapid exchange of nutrients. For example, work on the vertebrate kidney has suggested that negative charges in the BM may or may not be responsible for the passage of proteins entering into the urine [51].

The outer surface of several tissues and organs, such as the arteries, the gonad, the gut, and abdominal muscle, as well as the outer covering of the digestive gland and heart, is covered by sheaths bathed in hemolymph. These are composed of multiple layers of collagen fibers containing a few, widely separated fibroblast-like cells. Given the tensile strength of collagen, the sheaths may provide the greatest stability to the enclosed organs, but at what cost to nutrient exchange?

In this study, the distinction between BMs and connective tissue sheaths in shrimp is solely based on morphology, including the positive staining with PAS and the ultra-structural demonstration of an electron-dense lamina densa. The use of these terms and a definitive distinction between BMs and connective tissue sheaths is based on the presumed nature of the contained collagen; type IV is the definitive component of lamina densa and hence BMs and the sheaths are composed of type I [21,44]. The biochemical nature of collagens in arthropods is best known for insects [52]; in crustacea the interest in collagen studies has been primarily on the abdominal muscle in relation to food quality [53]. Type I, IV, and other types of collagen have been identified from both taxa [52–55] but immuno-localization of these proteins seems limited to work in insects [56]. Although collagens, such as type I and type IV, are highly conserved throughout animal evolution [31,57], commercially available, non-crustacean antibodies may not prove effective in verifying different types of collagen lining different tissues in the shrimp.

The demonstration of fibrillin and two types of collagens lining the hemal spaces complicates our understanding of the physiology of blood flow in crustaceans and presumably other invertebrates with an open circulatory system. In vertebrates, blood is confined within vessels lined by an endothelium. Contact of circulating hemocytes with collagen, such as the BM beneath the endothelium, indicates damage to the endothelium and is a trigger for blood clotting [58]. Clearly, neither collagen nor fibrillin are a trigger for blood coagulation in crustaceans [59]. In addition, the irregular surfaces of the connective tissue sheaths must produce a non-laminar flow, which in vertebrates also encourages blood coagulation [60,61].

It is hoped that this morphological study will encourage future studies to address three topics. First, the types of collagen in the layers need to be established. Second, the effect of the different layers on nutrient exchanges should be addressed. Third, what cells are producing these layers? Embedded cells, such as fibroblast-like cells and muscle, may be the source but seem too few and widely dispersed to cover the large surface area of the entire hemal space. It has been suggested, at least in insects [62–67], that these matrices are deposited by the circulating hemocytes, which could also be the evolutionary source of the endothelia seen in closed circulatory systems [4,68,69]. If this can be shown for crustaceans, then the role of the hemocytes will be greatly expanded from their primary research focus on innate immune responses and wound healing [7,10,70].

Author Contributions: G.M., R.B.S., S.B.: investigation; G.M.: decisions on methodology, collection and processing of samples, image analysis, and preparation of the manuscript. All authors have read and agreed to the published version of the manuscript.

Funding: This research received no external funding. R.B.S. received support from the Undergraduate Research Center at Occidental College.

Institutional Review Board Statement: Not applicable.

Informed Consent Statement: Not applicable.

Acknowledgments: We wish to thank Jonathan Williams and Dan Pondella of the Vantuna Research Group at Occidental College for catching the shrimp. We also wish to thank the Undergraduate Research Center at Occidental College, for financial assistance to R.B.S.

Conflicts of Interest: The authors declare no conflict of interest.

References

1. McGaw, I.J. The decapod crustacean circulatory system: A case that is neither open nor closed. *Microsc. Microanal.* **2005**, *11*, 18–36. [CrossRef]
2. Reiber, C.L.; McGaw, I.J. A review of the "open" and "closed" circulatory systems: New terminology for complex invertebrate circulatory systems in light of current findings. *Int. J. Zool.* **2009**, *2009*, e301284. [CrossRef]
3. McMahon, B.R. Control of cardiovascular function and its evolution in Crustacea. *J. Exp. Biol.* **2001**, *204*, 923–932. [CrossRef] [PubMed]
4. Monahan-Earley, R.; Dvorak, A.M.; Aird, W.C. Evolutionary origins of the blood vascular system and endothelium. *J. Thromb. Haemost.* **2013**, *11*, 46–66. [CrossRef] [PubMed]
5. Salt, G. *The Cellular Defense Reactions of Insects*; Cambridge University Press: Cambridge, UK, 1970. [CrossRef]
6. Rizki, T.M.; Rizki, R.M. The Cellular Defense System of Drosophila Melanogaster. In *Insect Ultrastructure*; King, R.C., Akai, H., Eds.; Springer: Boston, MA, USA, 1984; Volume 2, pp. 579–604. [CrossRef]
7. Loker, E.S.; Adema, C.M.; Zhang, S.M.; Kepler, T.B. Invertebrate immune systems—Not homogeneous, not simple, not well understood. *Immunol. Rev.* **2004**, *198*, 10–24. [CrossRef] [PubMed]
8. Yang, I.A.; Fong, K.M.; Holgate, S.T.; Holloway, J.W. The role of Toll-like receptors and related receptors of the innate immune system in asthma. *Curr. Opin. Allergy Clin. Immunol.* **2006**, *6*, 23–28. [CrossRef]
9. Davids, B.J.; Yoshino, T.P. Integrin-like RGD-dependent binding mechanism involved in the spreading response of circulating molluscan phagocytes. *Dev. Comp. Immunol.* **1998**, *22*, 39–53. [CrossRef]
10. Holmblad, T.; Söderhäll, K. Cell adhesion molecules and antioxidative enzymes in a crustacean, possible role in immunity. *Aquaculture* **1999**, *172*, 111–123. [CrossRef]
11. Plows, L.D.; Cook, R.T.; Davies, A.J.; Walker, A.J. Integrin engagement modulates the phosphorylation of focal adhesion kinase, phagocytosis, and cell spreading in molluscan defense cells. *Biochim. Biophys. Acta (BBA) Mol. Cell Res.* **2006**, *1763*, 779–786. [CrossRef]
12. Kawasaki, M.; Delamare-Deboutteville, J.; Dang, C.; Barnes, A.C. Hemiuroid trematode sporocysts are undetected by hemocytes of their intermediate host, the ark cockle Anadara trapezia: Potential role of surface carbohydrates in successful parasitism. *Fish Shellfish Immunol.* **2013**, *35*, 1937–1947. [CrossRef]
13. Chen, J.H.; Bayne, C.J. Bivalve mollusc hemocyte behaviors: Characterization of hemocyte aggregation and adhesion and their inhibition in the California Mussel (*Mytilus californianus*). *Biol. Bull.* **1995**, *188*, 255–266. [CrossRef]
14. Oka, M. Studies on Penaeus orientalis Kishinouye—VIII Structure of the newly found lymphoid organ. *Bull. Jpn. Soc. Sci. Fisher* **1969**, *35*, 245–250. [CrossRef]
15. Van de Braak, C.B.T.; Botterblom, M.H.A.; Liu, W.; Taverne, N.; van der Knaap, W.P.W.; Rombout, J.H.W.M. The role of the haematopoietic tissue in haemocyte production and maturation in the black tiger shrimp (Penaeus monodon). *Fish Shellfish Immunol.* **2002**, *12*, 253–272. [CrossRef] [PubMed]
16. Martin, G.G.; Hose, J.E.; Kim, J.J. Structure of hematopoietic nodules in the ridgeback prawn, Sicyonia ingentis: Light and electron microscopic observations. *J. Morphol.* **1987**, *192*, 204. [CrossRef]
17. Martin, G.G.; Hose, J.E.; Corzine, C.J. Morphological comparison of major arteries in the ridgeback prawn, Sicyonia ingentis. *J. Morphol.* **1989**, *200*, 175–183. [CrossRef]
18. Yurchenco, P.D.; Schittny, J.C. Molecular architecture of basement membranes. *FASEB J.* **1990**, *4*, 1577–1590. [CrossRef]
19. Asem, E.K.; Feng, S.; Stingley-Salazar, S.; Turek, J.; Peter, A.; Robinson, P. Basal lamina of avian ovarian follicle: Influence on morphology of granulosa cells in-vitro. *Comp. Biochem. Physiol. Part C Pharmacol. Toxicol. Endocrinol.* **2000**, *125*, 189–201. [CrossRef]
20. LeBleu, V.S.; MacDonald, B.; Kalluri, R. Structure and function of basement membranes. *Exp. Biol. Med.* **2007**, *232*, 1121–1129. [CrossRef]
21. Pedersen, K.J. Invited Review: Structure and composition of basement membranes and other basal matrix systems in selected invertebrates. *Acta Zool.* **1991**, *72*, 181–201. [CrossRef]
22. Timpl, R.; Dziadek, M. Structure, development, and molecular pathology of basement membranes. *Int. Rev. Exp. Pathol.* **1986**, *29*, 1–112. [PubMed]
23. Timpl, R. Structure and biological activity of basement membrane proteins. *Eur. J. Biochem.* **1989**, *180*, 487–502. [CrossRef]
24. Fidler, A.L.; Darris, C.E.; Chetyrkin, S.V.; Pedchenko, V.K.; Boudko, S.P.; Brown, K.L.; Jerome, W.G.; Hudson, J.K.; Roka, A.; Hudson, B.G. Collagen IV and basement membrane at the evolutionary dawn of metazoan tissues. *eLife* **2017**, *6*, e24176. [CrossRef] [PubMed]
25. Haraida, S.; Nerlich, A.G.; Wiest, I.; Schleicher, E.; Löhrs, U. Distribution of basement membrane components in normal adipose tissue and in benign and malignant tumors of lipomatous origin. *Mod. Pathol.* **1996**, *9*, 137–144.
26. Sanes, J.R. The basement membrane/basal lamina of skeletal muscle. *J. Biol. Chem.* **2003**, *278*, 12601–12604. [CrossRef]
27. Bruggink, A.H.; van Oosterhout, M.; de Jonge, N.; Cleutjens, J.P.M.; van Wichen, D.F.; van Kuik, J.; Tilanus, M.G.J.; Gmelig-Meyling, F.H.J.; van den Tweel, J.G.; de Weger, R.A. Type IV collagen degradation in the myocardial basement membrane after unloading of the failing heart by a left ventricular assist device. *Lab. Invest.* **2007**, *87*, 1125–1137. [CrossRef]
28. Yang, H.; Borg, T.K.; Wang, Z.; Ma, Z.; Gao, B.Z. Role of the basement membrane in regulation of cardiac electrical properties. *Ann. Biomed. Eng.* **2014**, *42*, 1148–1157. [CrossRef]

29. Grzelkowska-Kowalczyk, K. The Importance of Extracellular Matrix in Skeletal Muscle Development and Function. In *Composition and Function of the Extracellular Matrix in the Human Body*; IntechOpen: London, UK, 2016; pp. 3–24. [CrossRef]

30. Reggio, S.; Rouault, C.; Poitou, C.; Bichet, J.B.; Prifti, E.; Bouillot, J.L.; Rizkalla, S.; Lacasa, D.; Tordjman, J.; Clèment, K. Increased basement membrane components in adipose tissue during obesity: Links with TGFβ and metabolic phenotypes. *J. Clin. Endocrinol. Metab.* **2016**, *101*, 2578–2587. [CrossRef] [PubMed]

31. Sekiguchi, R.; Yamada, K.M. Basement membranes in development and disease. *Curr. Top. Dev. Biol.* **2018**, *130*, 143–191. [CrossRef] [PubMed]

32. Bussiere, C.T.; Wright, G.M.; DeMont, M.E. The mechanical function and structure of aortic microfibrils in the lobster Homarus americanus. *Comp. Biochem. Physiol. Part A Mol. Integr. Physiol.* **2006**, *143*, 417–428. [CrossRef]

33. Davison, I.G.; Wright, G.M.; DeMont, M.E. The structure and physical properties of invertebrate and primitive vertebrate arteries. *J. Exp. Biol.* **1995**, *198*, 2185–2196. [CrossRef] [PubMed]

34. McConnell, C.J.; Wright, G.M.; DeMont, M.E. The modulus of elasticity of lobster aorta microfibrils. *Experientia* **1996**, *52*, 918–921. [CrossRef]

35. Piha-Gossack, A.; Sossin, W.; Reinhardt, D.P. The evolution of extracellular fibrillins and their functional domains. *PLoS ONE.* **2012**, *7*, e33560. [CrossRef]

36. Kielty, C.M.; Baldock, C.; Lee, D.; Rock, M.J.; Ashworth, J.L.; Shuttleworth, C.A. Fibrillin: From microfibril assembly to biomechanical function. *Philos. Trans. R. Soc. Lond. Ser. B Biol. Sci.* **2002**, *357*, 207–217. [CrossRef]

37. Ramirez, F.; Sakai, L.Y.; Dietz, H.C.; Rifkin, D.B. Fibrillin microfibrils: Multipurpose extracellular networks in organismal physiology. *Physiol. Genom.* **2004**, *19*, 151–154. [CrossRef] [PubMed]

38. Jensen, S.A.; Robertson, I.B.; Handford, P.A. Dissecting the fibrillin microfibril: Structural insights into organization and function. *Structure* **2012**, *20*, 215–225. [CrossRef]

39. Chan, K.S.; Cavey, M.L.; Wilkens, J.L. Microscopic anatomy of the thin-walled vessels leaving the heart of the lobster Homarus americanus: Anterior lateral arteries. *Invertebr. Biol.* **2006**, *125*, 70–82. [CrossRef]

40. Factor, J.R.; Naar, M. The digestive system of the lobster, Homarus americanus: I. Connective tissue of the digestive gland. *J. Morphol.* **1985**, *184*, 311–321. [CrossRef] [PubMed]

41. Feng, T.P.; Liu, Y.M. The connective tissue sheath of the nerve as effective diffusion barrier. *J. Cell. Comp. Physiol.* **1949**, *34*, 1–16. [CrossRef] [PubMed]

42. Krnjević, K. The connective tissue of the frog sciatic nerve. *Q. J. Exp. Physiol. Cogn. Med. Sci. Transl. Integr.* **1954**, *39*, 55–72. [CrossRef] [PubMed]

43. Twarog, B.M.; Roeder, K.D. Properties of the connective tissue sheath of the cockroach abdominal nerve cord. *Biol. Bull.* **1956**, *111*, 278–286. [CrossRef]

44. Ashhurst, D.E.; Chapman, J.A. The tissue sheath of the nervous system of Locusta migratoria: An electron microscope study. *J. Cell Sci.* **1961**, *3*, 463–467. [CrossRef]

45. Anderson, S.L. Multiple spawning and molt synchrony in a free spawning shrimp (*Sicyonia ingentis*: Penaeoidea). *Biol. Bull.* **1985**, *168*, 377–394. [CrossRef]

46. Humason, G.L. *Animal Tissue Techniques*, 4th ed.; WH Freeman and Co.: San Francisco, CA, USA, 1979; p. 641.

47. Spurr, A.R. A low-viscosity epoxy resin embedding medium for electron microscopy. *J. Ultrastruct. Res.* **1969**, *26*, 31–43. [CrossRef]

48. Talbot, P. The ovary of the lobster, Homarus americanus. I. Architecture of the mature ovary. *J. Ultrastruct. Res.* **1981**, *76*, 235–248. [CrossRef]

49. Odselius, R.; Elofsson, R. The basement membrane of the insect and crustacean compound eye: Definition, fine structure, and comparative morphology. *Cell Tissue Res.* **1981**, *216*, 205–214. [CrossRef]

50. Nakao, T. The fine structure and innervation of gill lamellae in Anodonta. *Cell Tissue Res.* **1975**, *157*, 239–254. [CrossRef]

51. Miner, J.H. Glomerular filtration: The charge debate charges ahead. *Kidney Int.* **2008**, *74*, 259–261. [CrossRef]

52. Ashhurst, D.E. Integument, Respiration, and Circulation. In *Comprehensive Insect Physiology Biochemistry and Pharmacology*; Kerkut, G.A., Gilbert, L.I., Eds.; Pergamon Press: Oxford, UK, 1985; Volume 3, pp. 249–287.

53. Sriket, P.; Benjakul, S.; Visessanguan, W.; Kijroongrojana, K. Comparative studies on chemical composition and thermal properties of black tiger shrimp (*Penaeus monodon*) and white shrimp (*Penaeus vannamei*) meats. *Food Chem.* **2007**, *103*, 1199–1207. [CrossRef]

54. Yoshinaka, R.; Mizuta, S.; Itoh, Y.; Sato, M. Two genetically distinct types of collagen in kuruma prawn Penaeus japonicus. *Comp. Biochem. Physiol. Part B Comp. Biochem.* **1990**, *96*, 451–456. [CrossRef]

55. Sivakumar, P.; Suguna, L.; Chandrakasan, G. Molecular species of collagen in the intramuscular connective tissues of the marine crab, Scylla serrata. *Comp. Biochem. Physiol. Part B Biochem. Mol. Biol.* **2000**, *125*, 555–562. [CrossRef]

56. Mirre, C.; Le Parco, Y.; Knibiehler, B. Collagen IV is present in the developing CNS during Drosophila neurogenesis. *J. Neurosci. Res.* **1992**, *31*, 146–155. [CrossRef] [PubMed]

57. Naba, A.; Clauser, K.R.; Ding, H.; Whittaker, C.A.; Carr, S.A.; Hynes, R.O. The extracellular matrix: Tools and insights for the 'omics' era. *Matrix Biol.* **2016**, *49*, 10–24. [CrossRef]

58. Yau, J.W.; Teoh, H.; Verma, S. Endothelial cell control of thrombosis. *BMC Cardiovasc. Disord.* **2015**, *15*, 1–11. [CrossRef]

59. Perdomo-Morales, R.; Montero-Alejo, V.; Perera, E. The clotting system in decapod crustaceans: History, current knowledge and what we need to know beyond the models. *Fish Shellfish Immunol.* **2019**, *84*, 204–212. [CrossRef]

60. Zucker, M.B.; Borrelli, J. Platelet clumping produced by connective tissue suspensions and by collagen. *Proc. Soc. Exp. Biol. Med.* **1962**, *109*, 779–787. [CrossRef]
61. Gimbrone, M.A.; Anderson, K.R.; Topper, J.N. The critical role of mechanical forces in blood vessel development, physiology and pathology. *J. Vasc. Surg.* **1999**, *29*, 1104–1151. [CrossRef]
62. Wigglesworth, V.B. Digestion and Nutrition. In *The Principles of Insect Physiology*; Springer: Dordrecht, The Netherlands, 1972; pp. 476–552. [CrossRef]
63. Ball, E.E.; de Couet, H.G.; Horn, P.L.; Quinn, J.M.A. Haemocytes secrete basement membrane components in embryonic locusts. *Development* **1987**, *99*, 255–259. [CrossRef]
64. Knibiehler, B.; Mirre, C.; Cecchini, J.P.; Le Parco, Y. Haemocytes accumulate collagen transcripts during Drosophila melanogaster metamorphosis. *Roux's Arch. Dev. Biol.* **1987**, *196*, 243–247. [CrossRef]
65. Uhrík, B.; Rýdlová, K.; Zacharová, D. The roles of haemocytes during degeneration and regeneration of crayfish muscle fibres. *Cell Tissue Res.* **1989**, *255*, 443–449. [CrossRef]
66. Adachi, T.; Tomita, M.; Yoshizato, K. Synthesis of prolyl 4-hydroxylase alpha subunit and type IV collagen in hemocytic granular cells of silkworm, Bombyx mori: Involvement of type IV collagen in self-defense reaction and metamorphosis. *Matrix Biol.* **2005**, *24*, 136–154. [CrossRef]
67. Franklin, B.M.; Maroudas, E.; Osborn, J.L. Sine-wave electrical stimulation initiates a voltage-gated potassium channel-dependent soft tissue response characterized by induction of hemocyte recruitment and collagen deposition. *Physiol. Rep.* **2016**, *4*, e12832. [CrossRef]
68. Muñoz-Chápuli, R.; Carmona, R.; Guadix, J.A.; Macías, D.; Pérez-Pomares, J.M. The origin of the endothelial cells: An evo-devo approach for the invertebrate/vertebrate transition of the circulatory system. *Evol. Dev.* **2005**, *7*, 351–358. [CrossRef]
69. Hartenstein, V.; Mandal, L. The blood/vascular system in a phylogenetic perspective. *BioEssays* **2006**, *28*, 1203–1210. [CrossRef]
70. Söderhäll, K.; Cerenius, L. Crustacean immunity. *Annu. Rev. Fish Dis.* **1992**, *2*, 3–23. [CrossRef]

Journal of
*Marine Science
and Engineering*

Article

Differentiation of Crystal Cells, Gravity-Sensing Cells in the Placozoan *Trichoplax adhaerens*

Tatiana D. Mayorova

Laboratory of Neurobiology, Structural Cell Biology Section, National Institute of Neurological Disorders and Stroke, National Institutes of Health, 49 Convent Drive, Bethesda, MD 20892, USA; tatiana.mayorova@nih.gov or mayorova@wsbs-msu.ru

Abstract: *Trichoplax adhaerens* are simple animals with no nervous system, muscles or body axis. Nevertheless, *Trichoplax* demonstrate complex behaviors, including responses to the direction of the gravity vector. They have only six somatic cell types, and one of them, crystal cells, has been implicated in gravity reception. Multiple crystal cells are scattered near the rim of the pancake-shaped animal; each contains a cup-shaped nucleus and an intracellular crystal, which aligns its position according to the gravity force. Little is known about the development of any cell type in *Trichoplax*, which, in the laboratory, propagate exclusively by binary fission. Electron and light microscopy were used to investigate the stages by which crystal cells develop their mature phenotypes and distributions. Nascent crystal cells, identified by their possession of a small crystal, were located farther from the rim than mature crystal cells, indicating that crystal cells undergo displacement during maturation. They were elongated in shape and their nucleus was rounded. The crystal develops inside a vacuole flanked by multiple mitochondria, which, perhaps, supply molecules needed for the biomineralization process underlying crystal formation. This research sheds light on the development of unique cells with internal biomineralization and poses questions for further research.

Keywords: *Trichoplax*; Placozoa; cell type evolution; gravireception; crystal cell; biomineralization; lithocyte

Citation: Mayorova, T.D. Differentiation of Crystal Cells, Gravity-Sensing Cells in the Placozoan *Trichoplax adhaerens*. *J. Mar. Sci. Eng.* **2021**, *9*, 1229. https://doi.org/10.3390/jmse9111229

Academic Editor: Alexandre Lobo-da-Cunha

Received: 9 October 2021
Accepted: 4 November 2021
Published: 6 November 2021

Publisher's Note: MDPI stays neutral with regard to jurisdictional claims in published maps and institutional affiliations.

1. Introduction

Trichoplax adhaerens (Schulze, 1883) belongs to phylum Placozoa, a group of unique marine animals, the appearance of which is strikingly different from animals in other phyla. They have small (about 1 mm in diameter), flat, pancake-shaped bodies that undergo amoeboid-like changes in outline [1–3]. They adhere to and crawl on substrates, propelled by monociliated cells on the ventral side [4]. They lack a morphological anterior–posterior body axis and can crawl in any direction [5]. Using chemotaxis, *Trichoplax* find their food—cyanobacteria and microalgae [6]. Unlike most other metazoans, *Trichoplax* digest their food externally, secreting digestive enzymes into the cleft between their body and the substrate [4]. According to the majority of phylogenetic studies, Placozoa are sister to the clade that includes Cnidaria and Bilateria [7–10]. Their simple body plans and lifestyles make them a useful model for research aimed at understanding metazoan evolution.

Trichoplax has six broadly defined somatic cell types [3]. The main components of the dorsal and ventral epithelia are monociliary dorsal and ventral epithelial cells, respectively [4,11,12]. The ventral epithelium also includes mucocytes, several subtypes of peptidergic secretory cells, and lipophils, involved in secretion of digestive enzymes (reviewed by [13]). Each of these cell types is arranged in a distinctive radial pattern [14,15]. Fiber cells, implicated in innate immunity, are arrayed in a layer between the dorsal and ventral epithelia [3,16,17]. Crystal cells, each containing a birefringent crystal, are confined to a narrow zone near the rim where the dorsal and ventral epithelia meet [3,18].

Although there is molecular evidence of sexual reproduction in a clade of placozoans collected from the Caribbean [19], placozoans maintained in culture propagate exclusively asexually. They divide frequently by binary fission or bud off spheric swarmers [5,20–23]. Analysis of cell proliferation shows large numbers of mitotic cells throughout the body [17,24]. However, no information is available about the generation and differentiation of any cell type. The goal of this research was to learn about the development of one of the somatic cell types, crystal cells.

Crystal cells have distinctive morphological characteristics, the most prominent of which is possession of an intracellular birefringent crystal composed of calcium carbonate in the form of aragonite [3,18]. Mitochondria are closely apposed to the crystal and the cytoplasm is clear and almost free of organelles. The nucleus is cup-shaped and aligned along the plasma membrane of one side of the cell, with the opening of the cup oriented toward the rim of the body. Crystal cells are thought to be gravity receptors because the crystal shifts in the direction of gravity [18]. *Trichoplax* typically have multiple (~75) crystal cells located around their circumference [3]. However, some individuals have few or no crystal cells. Animals with typical numbers of crystal cells were able to maintain normal moving behavior, a random walk, when on a vertical substrate, whereas animals with few or no crystal cells drifted downwards, suggesting that crystal cells are needed to receive and respond to information about body orientation [18]. Whether and how crystal cells communicate with the ciliated cells that mediate crawling is unknown. Crystal cells are reported to generate action potentials [25] but no synaptic connections between crystal cells and other cell types are apparent in ultrastructural studies. Better understanding of *Trichoplax* gravireception may come from promising new experimental setups testing behavior and gene expression in different gravity conditions [26].

Crystal cells resemble lithocytes, gravity detectors found in only a few other animal groups (i.e., Acoela, flatworms) [27,28]. Ultrastructural studies of lithocytes have shed light on the organization of statocysts and gravity reception in these groups. However, little information is available about the development of lithocytes or the chemical bases of biomineralization in these cells. The present work uses electron and light microscopy to reveal the stages by which crystal cells, lithocytes of *Trichoplax*, acquire their distinctive mature morphotype and distribution. Histochemical stains for molecules implicated in extracellular calcium carbonate biomineralization in other animals failed to stain crystal cells, suggesting that the formation of their aragonite crystals depends on different biochemical pathways.

2. Materials and Methods

Animal care. *Trichoplax adhaerens* culture of the Grell (1971) strain (a gift from Leo Buss, Yale University) was maintained in artificial seawater (ASW) (Instant Ocean, Blacksburg, VA, USA) at room temperature with 1% Micro Algae Grow (Florida Aqua Farms, Dade City, FL, USA) as described earlier [29]. *Trichoplax* medium was partially refreshed once a week. Animals were fed red algae (*Rhodamonas salina*, Provasoli-Guillard National Center for Culture of Marine Plankton, East Boothbay, ME, USA).

Transmission electron microscopy. *Trichoplax* were frozen and freeze substituted as described elsewhere [3]. Samples were embedded in Epon resin and ultrathin sections were cut in series of 5–7 continuous sections in order to examine the one with the largest void from a crystal. After a short counterstaining with uranyl acetate and lead citrate, sections were examined at 80 or 120 KV in a JEOL 200-CX electron microscope (JEOL, Tokyo, Japan) and were photographed with an AMT camera mounted below the column.

Light microscopy. *Trichoplax* were placed on salinized cover slips and fixed in a mixture of 4% paraformaldehyde (EMS, Hatfield, PA, USA) and 0.25% glutaraldehyde (EMS, USA) in buffered ASW (NaCl 400 mM; $MgCl_2$ 5 mM; $CaCl_2$ 2 mM; sucrose 300 mM; HEPES 30 mM; pH 7.4) for two hours.

For crystal size evaluation, the samples were rinsed from fixative cocktail with PBS (pH 7.4), stained with Hoechst dye (1:2000; Life Technologies, Carlsbad, CA, USA), mounted in Vectashield (Vector Laboratories, Burlingame, CA, USA), and examined in a LSM 510 microscope (Carl Zeiss Microscopy, Thornwood, NY, USA). Images were captured with DIC and fluorescence optics with 405 nm illumination using 40X NA 1.2 or 63X NA1.4 objectives.

For labeling with different dyes, samples were rinsed in PBS and decalcified in 10% EDTA overnight. The following combinations of dyes were applied after rinsing with PBS with 0.2% Tween (PBST):

1. Calcofluor White (Sigma-Aldrich; St. Louis, MO, USA) diluted 1:1000 in PBST was applied for two hours; after washing with PBS, propidium iodide was added for 20 min.
2. Samples were incubated in blocking buffer (3% normal goat serum, 2% horse serum, 1% BSA in PBS) for one hour and then Col-F dye (20 µM final concentration; Immuno-Chemistry Technologies, Bloomington, MN, USA) was applied for 30 min at 37 °C. After washing with PBS, Hoechst was added for 20 min.
3. A cocktail of seven fluorescein conjugated lectins (Concanavalin A, *Dolichos biflorus* agglutinin, peanut agglutinin, *Ricinus communis* agglutinin, soybean agglutinin, *Ulex europaeus* agglutinin, and wheat germ agglutinin, all from Vector Laboratories, USA) each diluted 1:200 in PBST and Hoechst was applied for two hours.

The samples were rinsed from dyes with PBS, mounted in Vectashield, and examined in a LSM 510 microscope. Crystal cells were identified by a cup-shaped nucleus. Stacks of images were obtained with 63X NA 1.4 objective and were then processed with Fiji software.

Crystal measurement and statistics. A series of optical sections through crystal cells were captured with DIC optics. For each cell, the optical section capturing the crystal in focus was selected for measurement. The longest diagonal of the bright, rectangular crystal and the distance between the crystal and the rim of the animal were measured with the straight-line tool of Fiji 2.0.0-rc-49/1.51d software (NIH, Bethesda, MD, USA).

To analyze correlation between crystal size and its position, statistics treatment was performed with Past 3.14 software (Natural History Museum, Oslo, Norway). Since the normality tests were not passed, Spearman's rank correlation coefficient was calculated. To exclude heteroscedasticity, a Breusch Pagan test was applied.

3. Results

The developmental sequence by which crystal cells acquire their mature phenotype and distribution was investigated by transmission electron microscopy of thin sections and light microscopy in wholemounts.

Morphology of differentiating crystal cells. Morphologically, commitment of a cell to differentiation into crystal cell can be detected when a small crystal begins to grow in a vacuole surrounded by mitochondria (Figure 1A). The crystal in a differentiating crystal cell is several times smaller than crystals typically observed in mature crystal cells (compare Figure 1A,C). The content of the vacuole bathing the growing crystal is homogeneous and electron lucent (Figure 1A, inset). The differentiating crystal cell is elongated and irregular in outline (Figure 1A). The nucleus is round, oval or bean-shaped (Figure 1A,A') and may not be flattened to the plasma membrane (Figure 1A). The cytoplasm contains ribosomes and organelles.

As differentiation progresses, the crystal becomes bigger and occupies more of the vacuole internal space (Figure 1B). The nucleus adheres to one side of the cell (Figure 1B,B') and gradually assumes the shape of a cup with thick walls (Figure 1 B,B'). The shape of the cell becomes oval and its outline is mostly smooth but bears small protrusions (Figure 1B).

In mature crystal cells, the crystal is tightly enclosed within the vacuole membrane (Figure 1C). The nucleus is flattened in cross section (Figure 1C,C') and closely aligned along the plasma membrane on one side of the cell (Figure 1C). The cytoplasm contains few ribosomes and organelles. The cell is round, and its surface is mainly smooth but has small, thin protrusions (Figure 1C). The cells have extensive areas of close apposition to one or more ciliated epithelial cells and also interact with fiber cell bodies and processes [18].

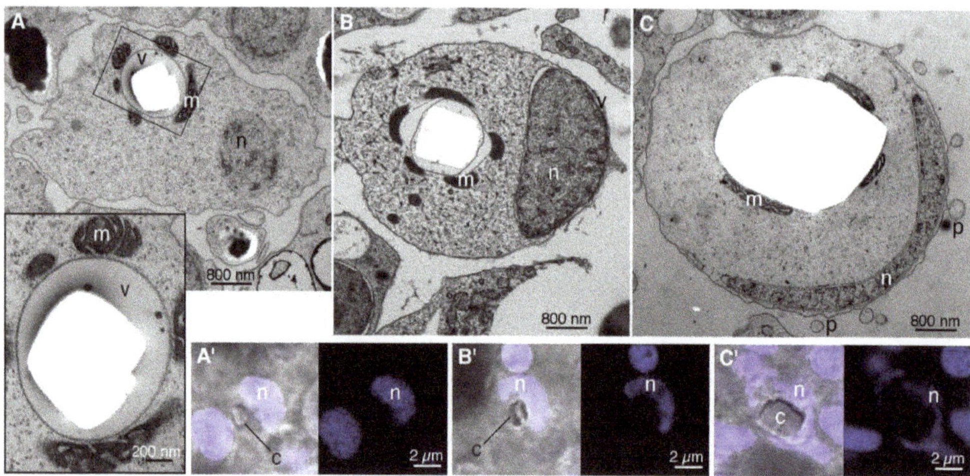

Figure 1. Transmission electron and light microscopy of developing and mature crystal cells. (**A–C**)—ultrathin sections in TEM. (**A′–C′**)—light microscopy of Hoechst labeling (blue) shown superimposed on DIC image (left) and separately (right). (**A**)—an early stage of crystal cell development characterized by a very small crystal (<800 nm), rounded nucleus, and elongated shape of the cell. Inset is a magnified boxed region, showing a vacuole with a growing crystal. (**A′**)—a crystal cell at the same stage as shown in (**A**), with a small crystal and bean-shaped nucleus. (**B**)—the crystal becomes larger, the nucleus adheres to the cell membrane and begins to adopt cup shape, and the cell rounds up. (**B′**)—a stage corresponding to (**B**) when the nucleus starts to bend. (**C**)—mature crystal cell with a large crystal occupying the entire volume of the vacuole and a thin cup-shaped nucleus. (**C′**)—mature crystal cell corresponding to that in (**C**) with a large crystal and a very thin nucleus. c—crystal, m—mitochondrion, n—crystal cell nucleus, p—crystal cell protrusions, v—vacuole with growing crystal.

The main morphological characteristics of a differentiating crystal cell are: (i) a smaller crystal floating inside a vacuole containing a substance that is electron lucent and homogeneous; (ii) a bean-shaped rather than cup-shaped nucleus and (iii) elongated rather than round shape of the cell. Acquisition of the mature phenotype involves growth of the crystal, reshaping of the nucleus and cell body and repositioning and removal of organelles.

Distribution of crystals of different size. Crystal cells are distributed along the rim of the animal and absent from central regions of the body [3,18]. Measurement of the distance from the body edge to individual crystals (n = 189) indicates that crystal cells are scattered within a concentric region starting approximately 15 μm from the edge and spreading inside up to 120 μm from the edge (Figure 2A). Within this belt-like zone, smaller crystals selectively occupy the deeper region, while large crystals are located closer to the periphery (Figure 2A,B). Indeed, there is a negative correlation between the size of a crystal and its distance from the rim of the body (Spearman's r_s = −0.55, p = 3.76 × 10^{-16}; homoscedasticity was confirmed with Breusch Pagan test giving a p value of 0.66).

Attempt to identify organic matrix for crystal biomineralization. To seek molecular signatures of an organic scaffold employed by crystal cells for biomineralization, several dyes were applied to decalcified tissues of *Trichoplax*: calcofluor white, a dye that binds chitin; Col-F dye, a stain for collagen and a cocktail of seven lectins with affinity to different sugar moieties. None of the dyes bind to material at sites of decalcified crystals (red arrowheads in Figure 3A–C), although some lectins label nuclear envelopes (Figure 3C).

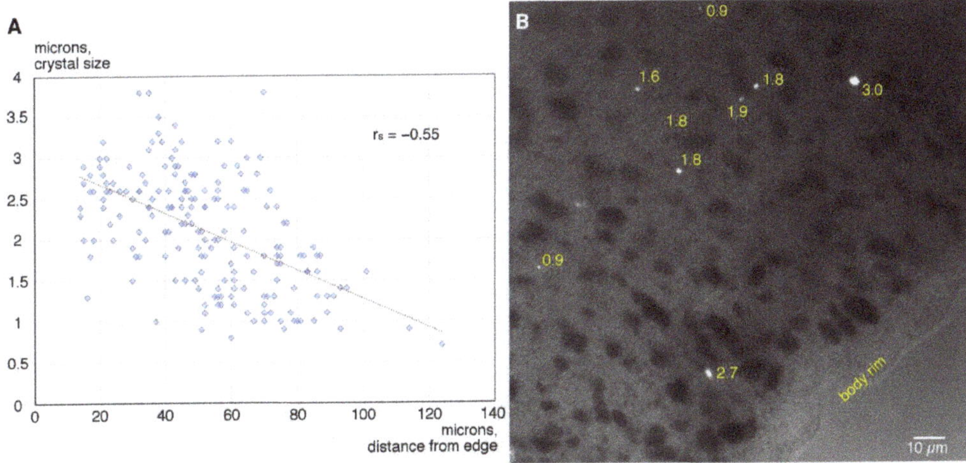

Figure 2. Distance from the rim of the animal of crystals of different size. (**A**)—dot plot showing individual crystal sizes at varying distances from the rim (blue semi-transparent diamonds; *N* crystals = 189; *N* animals = 8) and linear trend line (red). (**B**)—DIC image of marginal zone of *Trichoplax* body showing a group of crystals; size of each crystal is indicated in microns. Smaller crystals lie deeper while larger crystals lie closer to the periphery.

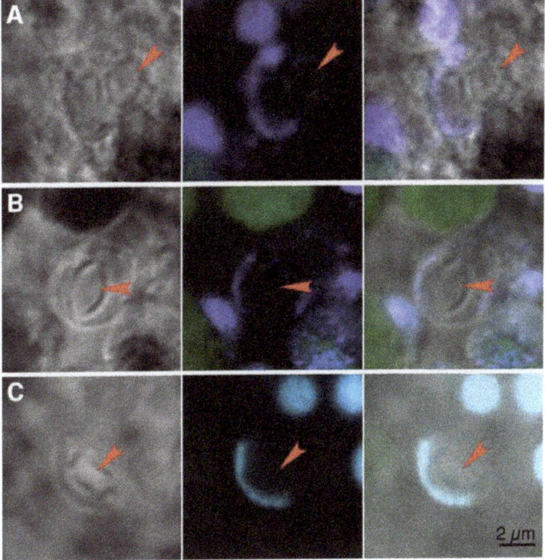

Figure 3. Fluorescence labeling (green) for potential components of organic matrix for crystal precipitation. Decalcified tissue was labeled with calcofluor white (**A**), Col-F (**B**) and a cocktail of lectins (**C**). Nuclear labeling (Propidium iodide in A and Hoechst in B and C) is blue. Left panel in each row shows an image in transmitted light, middle panel is merged fluorescence channels (maximum intensity projection through crystal cell thickness) and right panel is fluorescence channels superimposed on the transmitted light image. Arrowheads point at the site of a decalcified crystal. The scale bar is applicable to all images.

4. Discussion

We describe stages in the differentiation of placozoan crystal cells, a cell type implicated in gravity reception. Maturation of crystal cells involves growth of a crystal within a membrane enclosed vacuole, dramatic changes in the shape and position of the nucleus, reshaping of the cell body and extension of small protrusions, repositioning and removal of organelles, and cell displacement towards the rim. During maturation, they likely also establish their interactions with fiber cells and rim epithelial cells as described previously [18]. Crystal cell differentiation apparently occurs constitutively because most animals have a row of cells containing large crystals around their entire circumference as well as cells containing smaller crystals and located further from the rim. Nascent crystal cells do not contain specialized organelles that are features of other somatic cell types (dorsal and ventral epithelial cells, fiber cells, lipophil cells, and gland cells) so we hypothesize that they are generated by undifferentiated mitotic cells, which are prevalent throughout the body [17,30].

Cells involved in biomineralization in many animal lineages precipitate hard material in extracellular space. Cells precipitating hard material internally are rare and present in phylogenetically distant animal groups. These include scleroblasts in some corals [31,32] and lithocytes in Acoela [27,33,34], Xenoturbella [28], Ctenophore [35] and echinoderms [36]. Some unicellular organisms precipitate hard material intracellularly that is subsequently deposited extracellularly [37]. The patchy phylogenetic distributions of animals having cells with intracellular biomineralization is indicative of independent evolution of this cell type; likewise, the cells providing gravireceptions are thought not to share a common evolutionary origin [38]. Even within monophyletic clades, lithocytes are thought to have evolved independently in different lineages [28]. A feature that is shared between all of these various cell types is that internal biomineralization takes place within an intracellular vacuole. The membrane of this vacuole is thought to have a rich repertoire of ion pumps, transporters and exchangers [31,37]. Some lithocytes have mitochondria adhered to the crystal [39,40]. In placozoan crystal cells, both immature and mature, mitochondria surround the vacuole enclosing the crystal ([3,18]; present study) suggesting that mitochondria could be essential for biomineralization process. Perhaps they supply carbonate ions, as was shown in other calcifying non-bilaterians: sponges [41] and corals [42].

Information about the differentiation of cells with internal mineralization is sparse. Lithocytes of juvenile acoels are reported to have statoliths that are smaller and softer than those in mature animals [34]. In *Trichoplax*, a small growing crystal in a differentiating crystal cell is already hard as is evident because the crystal falls out during thin sectioning.

In most biomineralizing animals, hard inorganic material precipitates on an organic scaffold which guides crystallization in order to finally form a structure with specific characteristics (shape, ductility, density, hardness, and others) (reviewed by [43]). This organic scaffold often has a complex architecture and is composed of different molecules such as polysaccharides (i.e., chitin), collagen, proteoglycans, and small acidic proteins [44–47]. Our histochemical attempts to detect the epitopes of some of these molecules (chitin, collagen, and carbohydrate residues of proteoglycans) in the organic compartment of a crystal failed to identify a possible scaffold. Genome analysis revealed a very low level of proteins with high rate of negatively charged amino acid residues in *Trichoplax* [42], and transcriptome analysis did not reveal homologs of small acidic proteins in placozoans [48]. As placozoans lack scaffolding molecules involved in biomineralization in other animals, they may use a novel substrate, or cope with mineralization without any organic matrix.

Crystals of smaller size, that is, belonging to immature crystal cells, lay deeper inside the animal body than the bigger crystals. This difference suggests that immature crystal cells, apart from changing morphology, experience spatial displacement towards the periphery during differentiation. The mechanism of displacement is unknown; EM observations of crystal cell processes showed no evidence of motility [18]. Alternatively, crystal growth and crystal cell differentiation could be impeded in crystal cells differentiating farther from the animal rim. In addition to crystal cells, several other cell types in *Trichoplax* have distinct concentric spatial distributions [14,15]. The control of differentiation and positional information may be provided by multiple transcriptional factors with distinct expression patterns across the *Trichoplax* body [49–51].

The present research provides the first description of how one of the somatic cell types in Placozoa acquires its distinctive phenotype. Crystal cell differentiation involves a complex sequence of structural changes and utilization of biochemical processes that result in the formation of a hard intravacuolar crystal that we assume is critical for its proposed function as a gravity receptor. Much remains to be learned about biomineralization in crystal cells as well as about the sensory transduction mechanisms and signaling pathways involved in gravireception. A promising approach would be to use RNAseq techniques to characterize expression profile in crystal cells. A single cell RNAseq study of *Trichoplax* revealed multiple cell clusters, many of which could be assigned to specific morphological cell types [52]. However, no cluster was associated with crystal cells, possibly because crystal cells are the least prevalent cell type, making clustering hard to define. For further studies, samples enriched in crystal cells should be obtained; this could be achieved by selective collection of crystal cells based on their distinctive features, such as the possession of a birefringent crystal, and perhaps lower buoyancy than other cell types.

Funding: The work was supported by the Intramural Research Program of the National Institute of Neurological Disorders and Stroke and the National Institutes of Health.

Institutional Review Board Statement: Not applicable.

Informed Consent Statement: Not applicable.

Data Availability Statement: All data generated during the research are shown in the manuscript.

Acknowledgments: I am grateful to Carolyn L. Smith and Thomas S. Reese (NINDS, NIH) for advice and help in preparing the manuscript and for English proofreading.

Conflicts of Interest: I declare that I have no competing interests.

References

1. Schulze, F.E. Trichoplax adhaerens, nov. gen., nov. spec. *Zool. Anz.* **1891**, *6*, 92–97.
2. Grell, K.G.; Ruthmann, A. Placozoa. In *Microscopic Anatomy of Invertebrates, Volume 2: Placozoa, Porifera, Cnidaria, and Ctenophora*; Harrison, F.W., Westfall, J.A., Eds.; Wiley-Liss: New York, NY, USA, 1991; pp. 13–27.
3. Smith, C.L.; Varoqueaux, F.; Kittelmann, M.; Azzam, R.N.; Cooper, B.; Winters, C.A.; Eitel, M.; Fasshauer, D.; Reese, T.S. Novel cell types, neurosecretory cells, and body plan of the early-diverging metazoan Trichoplax adhaerens. *Curr. Biol.* **2014**, *24*, 1565–1572. [CrossRef]
4. Smith, C.L.; Pivovarova, N.; Reese, T.S. Coordinated feeding behavior in Trichoplax, an animal without synapses. *PLoS ONE* **2015**, *10*, e00822-17. [CrossRef]
5. Schierwater, B.; DeSalle, R. Placozoa. *Curr. Biol.* **2018**, *28*, R97–R98. [CrossRef]
6. Smith, C.L.; Reese, T.S.; Govezensky, T.; Barrio, R.A. Coherent directed movement toward food modeled in Trichoplax, a ciliated animal lacking a nervous system. *Proc. Natl. Acad. Sci. USA* **2019**, *116*, 8901–8908. [CrossRef]
7. Dunn, C.W.; Hejnol, A.; Matus, D.Q.; Pang, K.; Browne, W.E.; Smith, S.A.; Seaver, E.; Rouse, G.W.; Obst, M.; Edgecombe, G.D.; et al. Broad phylogenomic sampling improves resolution of the animal tree of life. *Nature* **2008**, *452*, 745–749. [CrossRef]
8. Moroz, L.L.; Kocot, K.M.; Citarella, M.R.; Dosung, S.; Norekian, T.P.; Povolotskaya, I.S.; Grigorenko, A.P.; Dailey, C.; Berezikov, E.; Buckley, K.M.; et al. The ctenophore genome and the evolutionary origins of neural systems. *Nature* **2014**, *510*, 109–114. [CrossRef]
9. Pett, W.; Adamski, M.; Adamska, M.; Francis, W.R.; Eitel, M.; Pisani, D.; Wörheide, G. The role of homology and orthology in the phylogenomic analysis of metazoan gene content. *Mol. Biol. Evol.* **2019**, *36*, 643–649. [CrossRef] [PubMed]
10. Simion, P.; Philippe, H.; Baurain, D.; Jager, M.; Richter, D.J.; Di Franco, A.; Roure, B.; Satoh, N.; Quéinnec, É.; Ereskovsky, A.; et al. A large and consistent phylogenomic dataset supports sponges as the sister group to all other animals. *Curr. Biol.* **2017**, *27*, 958–967. [CrossRef] [PubMed]
11. Ruthmann, A.; Behrendt, G.; Wahl, R. The ventral epithelium of Trichoplax adhaerens (Placozoa): Cytoskeletal structures, cell contacts and endocytosis. *Zoomorphology* **1986**, *106*, 115–122. [CrossRef]
12. Smith, C.L.; Reese, T.S. Adherens junctions modulate diffusion between epithelial cells in Trichoplax adhaerens. *Biol. Bull.* **2016**, *231*, 216–224. [CrossRef]
13. Smith, C.L.; Mayorova, T.D. Insights into the evolution of digestive systems from studies of Trichoplax adhaerens. *Cell Tissue Res.* **2019**, *377*, 353–367. [CrossRef]

14. Mayorova, T.D.; Hammar, K.; Winters, C.A.; Reese, T.S.; Smith, C.L. The ventral epithelium of Trichoplax adhaerens deploys in distinct patterns cells that secrete digestive enzymes, mucus or diverse neuropeptides. *Biol. Open* **2019**, *8*, bio045674. [CrossRef]

15. Varoqueaux, F.; Williams, E.A.; Grandemange, S.; Truscello, L.; Kamm, K.; Schierwater, B.; Jékely, G.; Fasshauer, D. High Cell Diversity and Complex Peptidergic Signaling Underlie Placozoan Behavior. *Curr. Biol.* **2018**, *28*, 3495–3501.e2. [CrossRef]

16. Rassat, J.; Ruthmann, A. Trichoplax adhaerens F.E. Schulze (placozoa) in the scanning electron microscope. *Zoomorphologie* **1979**, *93*, 59–72. [CrossRef]

17. Mayorova, T.D.; Hammar, K.; Jung, J.H.; Aronova, M.A.; Zhang, G.; Winters, C.A.; Reese, T.S.; Smith, C.L. Placozoan fiber cells: Mediators of innate immunity and participants in wound healing. *Nat. Sci. Rep.* **2021**, Submitted.

18. Mayorova, T.D.; Smith, C.L.; Hammar, K.; Winters, C.A.; Pivovarova, N.B.; Aronova, M.A.; Leapman, R.D.; Reese, T.S. Cells containing aragonite crystals mediate responses to gravity in Trichoplax adhaerens (Placozoa), an animal lacking neurons and synapses. *PLoS ONE* **2018**, *13*, e0190905. [CrossRef] [PubMed]

19. Signorovitch, A.Y.; Dellaporta, S.L.; Buss, L.W. Molecular signatures for sex in the Placozoa. *Proc. Natl. Acad. Sci. USA* **2005**, *102*, 15518–15522. [CrossRef]

20. Schierwater, B. My favorite animal, Trichoplax adhaerens. *BioEssays* **2005**, *27*, 1294–1302. [CrossRef] [PubMed]

21. Srivastava, M.; Begovic, E.; Chapman, J.; Putnam, N.H.; Hellsten, U.; Kawashima, T.; Kuo, A.; Mitros, T.; Salamov, A.; Carpenter, M.L.; et al. The Trichoplax genome and the nature of placozoans. *Nature* **2008**, *454*, 955–960. [CrossRef] [PubMed]

22. Zuccolotto-Arellano, J.; Cuervo-González, R. Binary fission in Trichoplax is orthogonal to the subsequent division plane. *Mech. Dev.* **2020**, *162*, 103608. [CrossRef]

23. Schierwater, B.; Osigus, H.J.; Bergmann, T.; Blackstone, N.W.; Hadrys, H.; Hauslage, J.; Humbert, P.O.; Kamm, K.; Kvansakul, M.; Wysocki, K.; et al. The enigmatic Placozoa part 1: Exploring evolutionary controversies and poor ecological knowledge. *BioEssays* **2021**, *43*, 2100080. [CrossRef] [PubMed]

24. Von Der Chevallerie, K.; Rolfes, S.; Schierwater, B. Inhibitors of the p53-Mdm2 interaction increase programmed cell death and produce abnormal phenotypes in the placozoon Trichoplax adhaerens (F.E. Schulze). *Dev. Genes Evol.* **2014**, *224*, 79–85. [CrossRef]

25. Romanova, D.Y.; Smirnov, I.V.; Nikitin, M.A.; Kohn, A.B.; Borman, A.I.; Malyshev, A.Y.; Balaban, P.M.; Moroz, L.L. Sodium action potentials in placozoa: Insights into behavioral integration and evolution of nerveless animals. *Biochem. Biophys. Res. Commun.* **2020**, *532*, 120–126. [CrossRef]

26. Schierwater, B.; Osigus, H.J.; Bergmann, T.; Blackstone, N.W.; Hadrys, H.; Hauslage, J.; Humbert, P.O.; Kamm, K.; Kvansakul, M.; Wysocki, K.; et al. The enigmatic Placozoa part 2: Exploring evolutionary controversies and promising questions on earth and in space. *BioEssays* **2021**, *43*, 2100083. [CrossRef] [PubMed]

27. Ferrero, E. A Fine Structural Analysis of the Statocyst in Turbellaria Acsela. *Zool. Scr.* **1973**, *2*, 5–16. [CrossRef]

28. Ehlers, U. Comparative morphology of statocysts in the Plathelminthes and the Xenoturbellida. *Hydrobiologia* **1991**, *227*, 263–271. [CrossRef]

29. Jackson, A.M.; Buss, L.W. Shiny spheres of placozoans (Trichoplax) function in anti-predator defense. *Invertebr. Biol.* **2009**, *128*, 205–212. [CrossRef]

30. Romanova, D.Y.; Varoqueaux, F.; Daraspe, J.; Nikitin, M.A.; Eitel, M.; Fasshauer, D.; Moroz, L.L. Hidden cell diversity in Placozoa: Ultrastructural insights from Hoilungia hongkongensis. *Cell Tissue Res.* **2021**, *385*, 623–637. [CrossRef]

31. Watabe, N.; Kingsley, R.J. Calcification in Octocorals. In *Hard Tissue Mineralization and Demineralization*; Suga, S., Watabe, N., Eds.; Springer: Tokyo, Japan, 1992; pp. 127–147.

32. Conci, N.; Vargas, S.; Wörheide, G. The Biology and Evolution of Calcite and Aragonite Mineralization in Octocorallia. *Front. Ecol. Evol.* **2021**, *9*, 623774. [CrossRef]

33. Achatz, J.G.; Martinez, P. The nervous system of Isodiametra pulchra (Acoela) with a discussion on the neuroanatomy of the Xenacoelomorpha and its evolutionary implications. *Front. Zool.* **2012**, *9*, 27. [CrossRef]

34. Sakagami, T.; Watanabe, K.; Ikeda, R.; Ando, M. Structural analysis of the statocyst and nervous system of Praesagittifera naikaiensis, an acoel flatworm, during development after hatching. *Zoomorphology* **2021**, *140*, 183–192. [CrossRef]

35. Tamm, S.L. Functional consequences of the asymmetric architecture of the ctenophore statocyst. *Biol. Bull.* **2015**, *229*, 173–184. [CrossRef]

36. Ehlers, U. Ultrastructure of the statocysts in the apodous sea cucumber Leptosynapta inhaerens (Holothuroidea, Echinodermata). *Acta Zool.* **1997**, *78*, 61–68. [CrossRef]

37. Taylor, A.R.; Chrachri, A.; Wheeler, G.; Goddard, H.; Brownlee, C. A voltage-gated H+ channel underlying pH homeostasis in calcifying Coccolithophores. *PLoS Biol.* **2011**, *9*, e1001085. [CrossRef] [PubMed]

38. Moroz, L.L. On the independent origins of complex brains and neurons. *Brain. Behav. Evol.* **2009**, *74*, 177–190. [CrossRef] [PubMed]

39. Ferrero, E.A.; Bedini, C. Ultrastructural aspects of nervous-system and statocyst morphogenesis during embryonic development of Convoluta psammophila (Turbellaria, Acoela). *Hydrobiologia* **1991**, *227*, 131–137. [CrossRef]

40. Rohde, K.; Watson, N.; Faubel, A. Ultrastructure of the Statocyst in an Undescribed Species of Luridae (Platyhelminthes, Rhabdocoela, Luridae). *Aust. J. Zool.* **1993**, *41*, 215–224. [CrossRef]

41. Voigt, O.; Adamski, M.; Sluzek, K.; Adamska, M. Calcareous sponge genomes reveal complex evolution of α-carbonic anhydrases and two key biomineralization enzymes. *BMC Evol. Biol.* **2014**, *14*, 1–19. [CrossRef]

42. Bhattacharya, D.; Agrawal, S.; Aranda, M.; Baumgarten, S.; Belcaid, M.; Drake, J.L.; Erwin, D.; Foret, S.; Gates, R.D.; Gruber, D.F.; et al. Comparative genomics explains the evolutionary success of reef-forming corals. *Elife* **2016**, *5*, e13288. [CrossRef] [PubMed]

43. Simkiss, K.; Wilbur, K.M. The Control of Mineralization. In *Biomineralization*; Academic Press Inc.: San-Diego, CA, USA, 1989; pp. 42–57.

44. Marin, F.; Luquet, G. Unusually Acidic Proteins in Biomineralization. In *Handbook of Biomineralization: Biological Aspects and Structure Formation*; Bäuerlein, E., Ed.; Wiley-VCH Verlag GmbH & Co. K: Hoboken, NJ, USA, 2007; pp. 273–290. ISBN 9783527316410.

45. Arias, J.L.; Fernández, M.S. Polysaccharides and proteoglycans in calcium carbonate-based Biomineralization. *Chem. Rev.* **2008**, *108*, 4475–4482. [CrossRef]

46. Ehrlich, H. Chitin and collagen as universal and alternative templates in biomineralization. *Int. Geol. Rev.* **2010**, *52*, 661–699. [CrossRef]

47. Nudelman, F.; Lausch, A.J.; Sommerdijk, N.A.J.M.; Sone, E.D. In vitro models of collagen biomineralization. *J. Struct. Biol.* **2013**, *183*, 258–269. [CrossRef]

48. Conci, N.; Wörheide, G.; Vargas, S. New non-bilaterian transcriptomes provide novel insights into the evolution of coral skeletomes. *Genome Biol. Evol.* **2019**, *11*, 3068–3081. [CrossRef] [PubMed]

49. Martinelli, C.; Spring, J. Distinct expression patterns of the two T-box homologues Brachyury and Tbx2/3 in the placozoan Trichoplax adhaerens. *Dev. Genes Evol.* **2003**, *213*, 492–499. [CrossRef]

50. Hadrys, T.; DeSalle, R.; Sagasser, S.; Fischer, N.; Schierwater, B. The trichoplax PaxB gene: A putative proto-Paxa/B/C gene predating the origin of nerve and sensory cells. *Mol. Biol. Evol.* **2005**, *22*, 1569–1578. [CrossRef] [PubMed]

51. Dubuc, T.Q.; Ryan, J.F.; Martindale, M.Q. "Dorsal-Ventral" Genes Are Part of an Ancient Axial Patterning System: Evidence from Trichoplax adhaerens (Placozoa). *Mol. Biol. Evol.* **2019**, *36*, 966–973. [CrossRef]

52. Sebé-Pedrós, A.; Chomsky, E.; Pang, K.; Lara-Astiaso, D.; Gaiti, F.; Mukamel, Z.; Amit, I.; Hejnol, A.; Degnan, B.M.; Tanay, A. Early metazoan cell type diversity and the evolution of multicellular gene regulation. *Nat. Ecol. Evol.* **2018**, *2*, 1176–1188. [CrossRef] [PubMed]

Journal of
*Marine Science
and Engineering*

Article

Functional Histology and Ultrastructure of the Digestive Tract in Two Species of Chitons (Mollusca, Polyplacophora)

Alexandre Lobo-da-Cunha [1,2,*], Ângela Alves [1,3], Elsa Oliveira [1,3] and Gonçalo Calado [4,5]

1   Department of Microscopy, Institute of Biomedical Sciences Abel Salazar (ICBAS), University of Porto, Rua Jorge Viterbo Ferreira 228, 4050-313 Porto, Portugal; aralves@icbas.up.pt (Â.A.); emoliveira@icbas.up.pt (E.O.)
2   Interdisciplinary Centre of Marine and Environmental Research (CIIMAR), 4450-208 Matosinhos, Portugal
3   Multidisciplinary Unit for Biomedical Research (UMIB), ICBAS, University of Porto, 4050-313 Porto, Portugal
4   Department of Life Sciences, Lusófona University, Campo Grande 376, 1749-024 Lisbon, Portugal; goncalo.calado@ulusofona.pt
5   MARE—Marine and Environmental Sciences Centre, NOVA School of Science and Technology (FCT NOVA), Campus de Caparica, 2829-516 Caparica, Portugal
*   Correspondence: alcunha@icbas.up.pt

Abstract: To continue the investigation on the digestive system of polyplacophoran molluscs, a histological and ultrastructural study of the oesophagus, stomach and intestine of *Chaetopleura angulata* and *Acanthochitona fascicularis* was carried out. Stomach content examination revealed an omnivorous diet. In both species the epithelium of the whole digestive tract consisted mostly of elongated absorptive cells with an apical border of microvilli. Cilia were also frequently present. Mitochondria and electron-dense lysosomes were the prominent organelles in the region above the nucleus. The basal region was characterised by an association of mitochondria, peroxisomes and lipid droplets. In general, glycogen deposits were also abundant in absorptive cells. The ultrastructural features indicate that the absorptive cells of the digestive tract epithelium are involved in endocytosis, intracellular digestion and storage of reserves. Histochemical techniques showed that the secretory cells of the digestive tract contained proteins and polysaccharides in their secretory vesicles. The secretory cells with vesicles of low electron density were classified as mucous cells, and the ones with electron-dense vesicles were designated basophilic cells due to their staining by basic dyes in light microscopy. Additionally, basal cells that seem to correspond to enteroendocrine cells containing oval electron-dense vesicles were found along the digestive tract epithelium of both species. The thin outer layer of the digestive tract wall consisted of muscle cells and nerves embedded in connective tissue.

Keywords: oesophagus; stomach; intestine; electron microscopy; histochemistry; Polyplacophora

Citation: Lobo-da-Cunha, A.; Alves, Â.; Oliveira, E.; Calado, G. Functional Histology and Ultrastructure of the Digestive Tract in Two Species of Chitons (Mollusca, Polyplacophora). *J. Mar. Sci. Eng.* **2022**, *10*, 160. https://doi.org/10.3390/jmse10020160

Academic Editor: Maria Gabriella Marin

Received: 20 December 2021
Accepted: 24 January 2022
Published: 26 January 2022

Publisher's Note: MDPI stays neutral with regard to jurisdictional claims in published maps and institutional affiliations.

1. Introduction

Polyplacophorans, commonly known as chitons, are benthic marine molluscs characterized by their eight articulated dorsal shell plates surrounded by a flexible marginal girdle, which allows them to curl into a ball. This class of molluscs comprises approximately a thousand extant species, many of them living on rocky shores and others in deeper ocean floors [1–4]. The digestive system of chitons was described in classical articles [5], but very few studies were specifically dedicated to this subject in the last decades [6–9]. Moreover, the ultrastructure of the digestive tract was not previously investigated in these molluscs.

Most chitons are grazers that scrape hard surfaces in order to collect the algae and invertebrates included in their diets [8,10–12]. Others are ambush predators [13] or feed on sunken wood [14]. The broad and long radula used for feeding contains numerous hard teeth mineralised with magnetite and hydroxyapatite. It is formed within the radular sac that can extend back from the buccal cavity to approximately one third of the animal length [1,15]. Beneath the radular sac lies the subradular sac with a mucus-secreting epithelium. A bilobed subradular organ located dorsally near the distal end of the subradular sac is lined by an epithelium mainly formed by cells with an apical microvillous border, containing electron-dense vesicles in the cytoplasm. Ciliated cells and some mucous cells are also present, and the dense innervation at the base of the epithelium suggests that the subradular organ has a chemosensory function [16].

Among chitons, a pair of small glands that have been called salivary glands are formed by large mucous cells interspersed with smaller wedge-shaped cells, but they were never studied by electron microscopy. In *Acanthochitona crinita*, *Cryptochiton stelleri* and *Stenoplax magdalenensis* these glands are branched and linked to the buccal cavity through very short ducts. Conversely, in *Lepidochitona cinerea*, a small size species, they are a simple sac opening directly into dorso-lateral buccal pouches [5]. The oesophagus is relatively short and it is linked to two large pouches known as "sugar glands" as they contain polysaccharide digesting enzymes [17]. These glands that were recently investigated by light and electron microscopy contain secretory cells and absorptive cells with many lysosomes [9]. The stomach of chitons is large and the gastric epithelium is formed by ciliated and non-ciliated supporting cells and secretory cells. The stomach is surrounded by branches of the digestive gland that is connected to the stomach by ducts [5,6]. Departing from the stomach, the long intestine is coiled around the digestive gland. A valve separating the anterior from the posterior intestine was reported in several species and the intestine ends in a short rectum with the anus located at the posterior end of the ventral surface of the animal [5,6,18]. However, despite some anatomical and histological descriptions, much remains to be known about the digestive tract of these marine molluscs. Thus, to continue the investigation on the digestive system of polyplacophorans, the oesophagus, stomach and intestine of *Chaetopleura angulata* (Spengler, 1797) and *Acanthochitona fascicularis* (Linnaeus, 1767) were investigated by light and transmission electron microscopy. Stomach content was also analysed to obtain some information about the diet of these species.

2. Materials and Methods

2.1. Collection Site, Digestive Tract Content and Measurements

Specimens of *Achantochitona fascicularis*, 3.0–4.5 cm in length, and *Chaetopleura angulata*, approximately 6 cm in length, were collected intertidally on Caldeira de Troia beach (38.49° N, 8.89° W) at the mouth of the Sado estuary (Portugal) (Figure S1). For histology and ultrastructure, live specimens were transported to the laboratory in a container with local water. Five specimens of each species were used for histological sections and small segments of the digestive tract, also obtained from 5 specimens, were processed for semithin and ultrathin sections. For analysis of stomach content, 3 specimens of each species were dissected at the collection site and their digestive tract was immediately preserved in 10% formalin made with seawater. In the laboratory the stomach content was analyzed and photographed under a stereomicroscope and further observations were made with a light microscope. For length determination, the intestine was carefully removed from unfixed specimens and measured with a ruler avoiding stretching. Several unfixed transverse sections of the anterior and posterior intestine without fecal material in the lumen were photographed under a stereomicroscope and the images were used to evaluate the intestine diameter. Other measurements were made on photomicrographs of semithin or ultrathin sections.

2.2. Tissue Processing for Light and Transmission Electron Microscopy

For histological sections, isolated oesophagus, whole stomachs and several segments of anterior and posterior intestine were fixed for 24 h with Bouin's solution, dehydrated in a graded series of ethanol and embedded in paraffin. Additionally, the same procedure was applied to the entire visceral mass of other specimens. Sections with a thickness of 5 μm were stained with Masson's trichrome, dehydrated, cleared with xylene and mounted with Coverquick 2000 medium.

For semithin and ultrathin sections, several samples of oesophagus, stomach, anterior and posterior intestine were fixed for 2 h at 4 °C in 2.5% glutaraldehyde and 4% formaldehyde (obtained from hydrolysis of para-formaldehyde), diluted with 0.4 M cacodylate buffer pH 7.4 (final buffer concentration 0.28 M). After washing in buffer, samples were postfixed with 2% osmium tetroxide buffered with cacodylate, dehydrated in increasing concentrations of ethanol and embedded in epoxy resin (Fluka). For light microscopy observations semithin sections with a thickness of 2 μm were stained with methylene blue and azure II. Semithin sections were also used for histochemical stainings. Ultrathin sections were stained with uranyl acetate for 20 min and lead citrate for 10 min. To enhance glycogen staining, some ultrathin sections were stained with 2% uranyl acetate for 10 min after 10 min treatment with a solution of tannic acid at 5% [19]. Ultrathin sections were observed in a JEOL 100CXII transmission electron microscope.

2.3. Histochemistry

Histological sections of Bouin-fixed samples with a thickness of 5 μm were stained by the periodic acid–Schiff (PAS) reaction for polysaccharide detection, Alcian blue staining at pH 1.0 for sulphated acid polysaccharides or Alcian blue at pH 2.5 for carboxylated acid polysaccharides. For PAS reaction sections were oxidised with 1% periodic acid for 10 min, washed with water, and stained with Schiff reagent for 20 min. For Alcian blue staining sections were stained for 30 min in 1% Alcian blue in a HCl solution with pH 1.0 or in acetic acid for pH 2.5 [20]. Some sections were stained by both PAS reaction and Alcian blue. Additionally, PAS reaction, Alcian blue staining, tetrazonium coupling reaction for protein detection and Sudan black staining for lipids were applied to 2 μm thick semithin sections of epoxy embedded samples. For PAS reaction, semithin sections with resin were treated with 1% periodic acid for 10 min, washed with water, and stained with Schiff reagent for 20 min. For Alcian blue staining, resin was removed using an alcoholic solution of sodium ethoxide, prepared by dissolving sodium hydroxide to saturation in absolute ethanol [21]. Subsequently, sections were thoroughly washed in absolute ethanol and water, and stained for 30 min in 1% Alcian blue solution in acetic acid, for pH 2.5, or in a HCl solution with pH 1.0. For the tetrazonium reaction, semithin sections with resin were treated with a freshly-prepared 0.2% solution of fast blue salt B in 0.1 M veronal-acetate buffer pH 9.2, for 10 min. After washing, sections were treated for 15 min with a saturated solution of β-naphthol in 0.1 M veronal-acetate buffer pH 9.2. For Sudan black staining semithin sections with resin were treated for 10 min with a 0.02% H_2O_2 solution to remove osmium tetroxide. After washing in water, sections were stained for 5 min with a saturated solution of Sudan black in 70% ethanol and washed in 70% ethanol. After washing, semithin sections with resin were air dried before being mounted with Coverquick 2000 medium. Alcian blue stained semithin sections without resin were dehydrated and cleared with xylene before mounting.

2.4. Ultrastructural Detection of Arylsulphatase Activity

For detection of the lysosomal enzyme arylsulphatase, small fragments of stomach and intestine were fixed for 1 h at 4 °C in 2.5% glutaraldehyde and 4% formaldehyde (obtained from hydrolysis of para-formaldehyde), diluted in 0.4 M cacodylate buffer pH 7.4. After washing in buffer, samples were incubated during 45 min at 35 °C in medium containing 40 mM $BaCl_2$ and 25 mM p-nitrocatechol sulfate in 0.2 M acetate buffer pH 5.0 [22]. For control, some tissue fragments were incubated in medium without p-nitrocatechol sulfate. Post-fixation was carried out for 2 h at room temperature with 1% OsO_4 and 1.5% potassium ferrocyanide in cacodylate buffer. Ultrathin sections were observed without further staining.

3. Results

3.1. General Morphology and Content of the Digestive Tract

A short oesophagus with an average diameter of 1.4–1.5 mm in *Chaetopleura angulata* and 0.9–1.0 mm in *Acanthochitona fascicularis* linked the buccal cavity to stomach. The large stomach with a complex shape presented a constriction in which a portion of the digestive gland was lodged. The long intestine departed from the stomach and formed several coils around the digestive gland (Figure 1A,B). In specimens of *C. angulata* approximately 6 cm long the total intestine length was 17.0–20.0 cm, being 7.0–10.5 cm in *A. fascicularis* specimens with a body length of 3.0–4.5 cm. Accordingly, in these species the intestine length/animal length ratio was in average 3.0 (SD 0.3) and 2.4 (SD 0.3), respectively. In *C. angulata* the diameter of the intestine was around 1.3–1.7 mm in the region close to the stomach and 0.8–1.2 mm for most of its length. In *A. fascicularis* a short constricted region with a length of 3.0–4.0 mm formed a valve separating the anterior from the posterior intestine (Figure 1A inset). In this last species, the anterior intestine corresponded to 20–25% of the total intestine length and had a diameter of 1.2–1.4 mm in the largest specimens that were observed, whereas in the posterior intestine the diameter was around 0.7–1.1 mm. In *C. angulata* a constricted region clearly separating the anterior from the posterior intestine was not observed, and in this case the first 3–4 cm of intestine (approximately 20% of the total length) were considered to correspond to the anterior intestine.

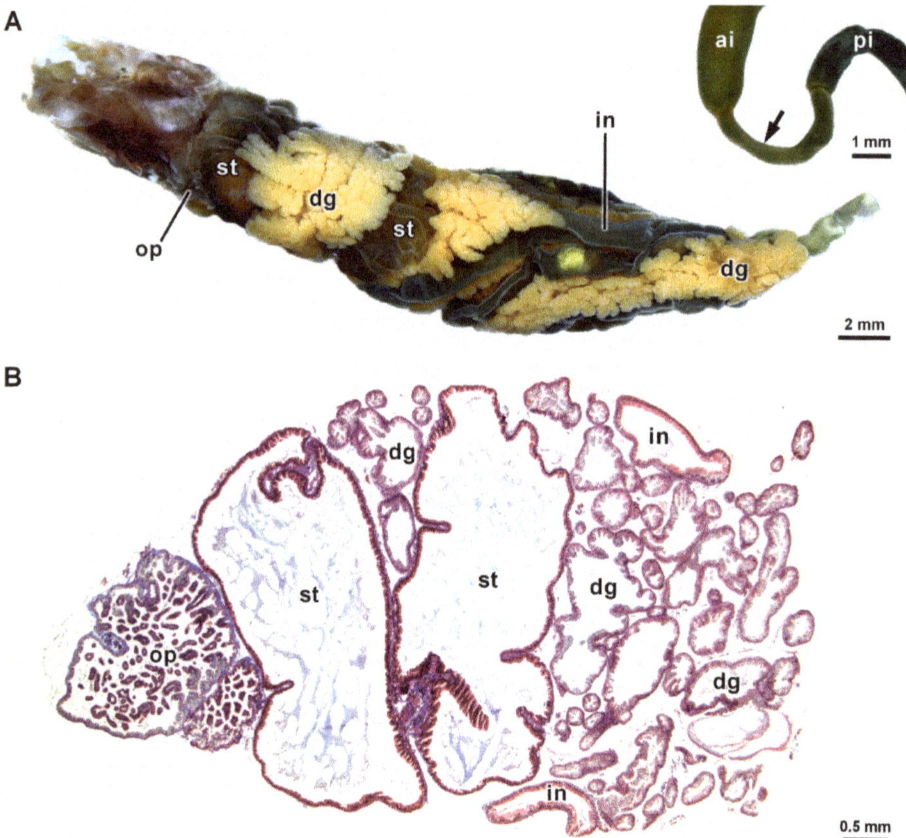

Figure 1. General view of the digestive system. (**A**) Stereomicroscope view of the digestive system of *A. fascicularis*. Inset, narrow segment (arrow) corresponding to the valve between the anterior (ai) and posterior intestine (pi). (**B**) Histological section of the digestive system of *C. angulata* stained by Masson's trichrome. dg—digestive gland, in—intestine, op—oesophageal pouches (sugar glands), st—stomach.

The analysis of the stomach content of *C. angulata* revealed remains of crustaceans, a few foraminiferans, diatoms and fragments of multicellular green and red algae

(Figure 2A–E). The stomach of *A. fascicularis* contained mainly fragments of multicellular green and red algae, diatoms and a few foraminiferans among unidentifiable debris (Figure 2F–I). However, small crustaceans or other invertebrates were not found in the digestive tract of *A. fascicularis*.

Figure 2. Digestive tract content. (**A–E**) A caprellid amphipod (**A**), fragments of other crustaceans (**B**), a piece of red algae (**C**), a diatom (**D**) and a foraminiferan (**E**) from the stomach of *C. angulata*. (**F–I**) Fragments of green (**F**) and red algae (**G**), a diatom frustule (**H**) and a foraminiferan (**I**) from the stomach of *A. fascicularis*.

3.2. Histology and Ultrastructure of the Oesophagus

The oesophagus had ridges created by large differences in epithelial thickness. The height of this epithelium was very variable, being around 25 µm at the thinner points between ridges and reaching about 200 µm in some ridges in transverse semithin sections of the oesophagus of both species. The outer layer of the oesophageal wall consisted of connective tissue, muscle cells and many nerves, usually with a thickness of 40–60 µm in *C. angulata* and 20–30 µm in *A. fascicularis* (Figure 3A–C). Light microscopy observations showed that the oesophageal epithelium consisted mostly of very thin and long absorptive cells with a microvilli border and cilia. In semithin sections stained by methylene blue and azure II, several blue stained bodies were detectable in the apical region of the absorptive cells. Osmiophilic lipid droplets were also noticeable in these cells (Figure 3D). The nuclei were elongated, several of them about 2.5 µm wide and 15 µm long, and in general were positioned in the central region of the cell.

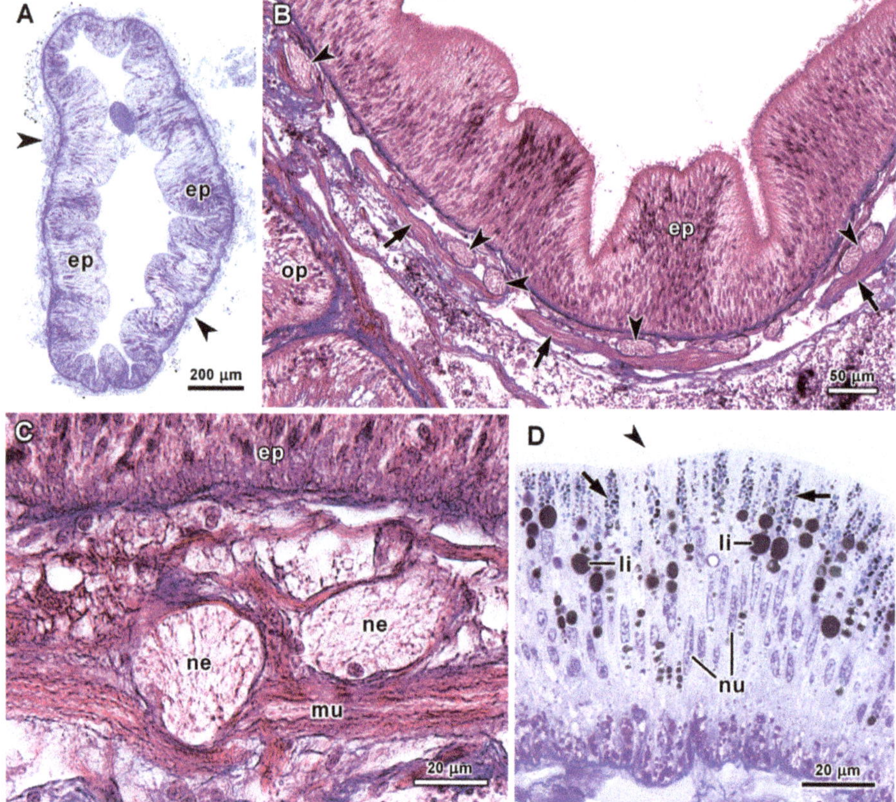

Figure 3. Histology of the oesophagus of *A. fascicularis* (**A**) and *C. angulata* (**B–D**). (**A**) Transverse semithin section of the oesophagus. The epithelium (ep) is surrounded by a layer of connective tissue with muscular cells and nerves (arrowheads). Methylene blue and azure II staining. (**B**) Several nerves (arrowheads) and muscle fibres (arrows) surround the oesophageal epithelium (ep). Masson's trichrome staining. (**C**) Higher magnification view of two nerves (ne) and muscle fibres (mu) below the oesophageal epithelium (ep). Masson's trichrome staining. (**D**) Absorptive cells with a microvilli border (arrowhead), inclusions in the apical region of the cytoplasm (arrows) and lipid droplets (li) in a semithin section of the oesophagus stained by methylene blue and azure II. nu—nuclei, op—oesophageal pouches (sugar glands).

Two types of secretory cells were identified in the oesophageal epithelium. In *A. fascicularis* the basophilic secretory vesicles of both cell types presented a dark blue colour in semithin sections stained by methylene blue and azure II. These secretory vesicles were also stained by tetrazonium and PAS reactions indicating a secretion rich in both proteins and polysaccharides (Figure 4A–E). However, in one secretory cell type the vesicles were significantly larger, most of them with a diameter of 1.0–2.5 μm in semithin sections, whereas in the other secretory cell type they seldom reached 0.9 μm (Table 1). Additionally, Alcian blue staining revealed the presence of acid polysaccharides only in the secretion of the cells with smaller vesicles (Figure 4F), which were designated as basophilic cells with small vesicles. For distinction, the secretory cells with larger vesicles were designated as basophilic cells with large vesicles (Table 1). In *C. angulata* oesophagus, in one cell type the cytoplasm above the nucleus was filled with a compact mass of secretory vesicles that stained purple-blue with methylene blue and azure II (Figure 4G). These secretory cells that were classified as mucous were stained by PAS reaction (Figure 4H) and Alcian blue at pH 1.0 and 2.5, but staining with the tetrazonium reaction was weaker (Table 1). The other type of secretory cells in *C. angulata* oesophagus was characterized by vesicles clearly separated from each other, reaching 3.0 μm in diameter in semithin sections. Nevertheless, some cells contained smaller secretory vesicles. These basophilic vesicles stained dark blue with methylene blue and azure II, and were strongly stained by PAS and tetrazonium reactions in semithin sections (Figure 4I), but were not stained by Alcian blue. At the bottom of the epithelium, basal cells were abundant with many small purple vesicles in semithin sections stained by methylene blue and azure II (Figure 4J), which were also stained by the tetrazonium coupling reaction (Table 1).

Table 1. Characterization of secretory cells of the digestive tract of *C. angulata* and *A. fascicularis*.

Organs	Species	Cell Types	Diameters of Secretory Vesicles (μm) [1]	PAS Reaction (Polysaccharides)	Alcian Blue pH 1.0 and 2.5 (Acid Polysaccharides)	Tetrazonium Reaction (Proteins)	Electron Density of Vesicles
Oesophagus	*C. angulata*	Basophilic	1.0–3.0	+ + +	−	+ + +	High
		Mucous	0.3–0.7	+ +	+ + +	+ +	Low
	A. fascicularis	Basophilic with small vesicles	0.4–0.9	+ + +	+ + +	+ +	High
		Basophilic with large vesicles	1.0–2.5	+ + +	−	+ + +	High
Stomach	*C. angulata*	Basophilic	0.4–0.8	+ + +	−	+ + +	High
	A. fascicularis	Basophilic	0.6–1.1	+ + +	−	+ + +	High
Intestine	*C. angulata*	Basophilic	0.4–0.8	+ + +	−	+ + +	High
		Mucous [2]	0.8–1.8	−	+ + +	−	Low or median
	A. fascicularis	Basophilic	0.5–1.5	+ + +	+ +	+ +	High
		Mucous [3]	(undetermined due to fusion)	+ + +	+ +	+	Low
	C. angulata	Basal cells (enteroendocrine)	0.4–1.0	−	−	+ + +	High
	A. fascicularis	Basal cells (enteroendocrine)	0.2–0.5	−	−	+ + +	High

Average stain intensity: − unstained; + weak; + + moderate; + + + strong; [1] Corresponding at least to 95% of the vesicle diameters in semithin or ultrathin sections; [2] In the valve between the anterior and posterior intestine, and in the posterior intestine; [3] In the valve between the anterior and posterior intestine.

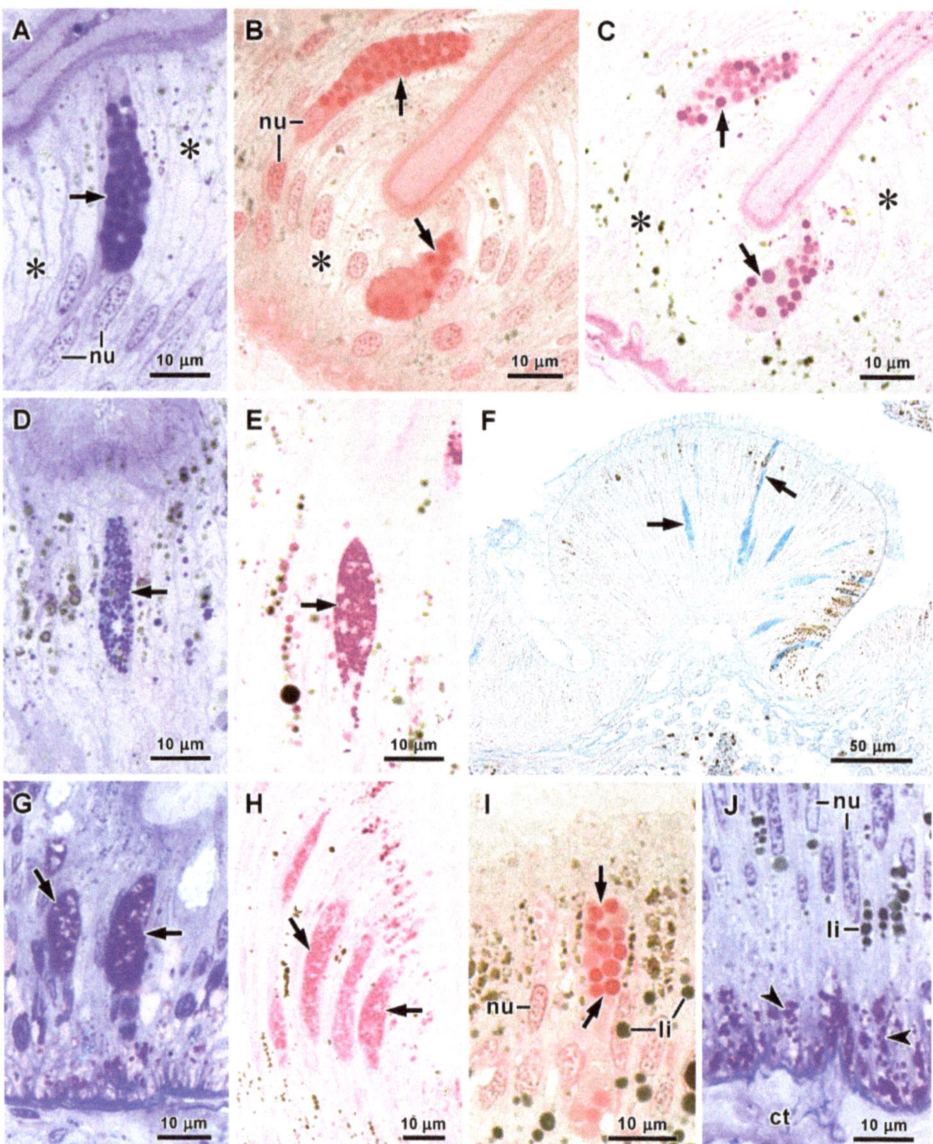

Figure 4. Oesophagus of *A. fascicularis* (**A–F**) and *C. angulata* (**G–J**), in semithin (**A–E,G–J**) and histological (**F**) sections. (**A–C**) Absorptive cells (asterisks) and basophilic cells with large secretory vesicles (arrows) in the oesophagus of *A. fascicularis* stained by methylene blue and azure II (**A**), tetrazonium coupling reaction for protein detection (**B**) and PAS reaction for polysaccharide detection (**C**). (**D–F**). Basophilic cells with small secretory vesicles (arrows) in the oesophagus of *A. fascicularis* stained by methylene blue and azure II (**D**), PAS reaction (**E**) and Alcian blue for acid polysaccharide detection (**F**). (**G,H**) Mucous cells (arrows) in the oesophagus of *C. angulata* stained by methylene blue and azure II (**G**) and PAS reaction (**H**). (**I**) Basophilic cell with large secretory vesicles (arrows) in the oesophagus of *C. angulata* stained by the tetrazonium coupling reaction. (**J**) Basal cells of the oesophagus with purple vesicles (arrowheads) in a section stained with methylene blue and azure II. ct—connective tissue, li—lipid droplets, nu—nuclei.

The absorptive ciliated cells of the oesophagus were ultrastructurally similar in both species. Several mitochondria, vesicles and some multivesicular bodies occurred in the most apical portion of the cytoplasm. In these cells, the electron-dense lysosomes were the prominent features in the cytoplasm above the elongated nucleus (Figure 5A). The vast majority of these lysosomes had diameters of 0.5–2.0 μm in ultrathin sections, and a few largest ones presented diameters up to 3.5 μm. These organelles correspond to the blue inclusions observed by light microscopy in semithin sections stained with methylene blue and azure II. Some mitochondria, lipid droplets, Golgi stacks and endoplasmic reticulum cisternae were present in the middle of these cells. Conversely, the most basal region was characterized by a large accumulation of mitochondria associated with peroxisomes and some lipid droplets (Figure 5B). Crystalline cores were absent in the peroxisomes that had diameters of 0.3–0.6 μm in ultrathin sections. Deep cell membrane invaginations were another feature of the basal region of absorptive cells. In the oesophagus of *C. angulata* the basophilic cells contained round electron-dense secretory vesicles and the flattened cisternae of rough endoplasmic reticulum filled a considerable part of the cytoplasm around the secretory vesicles (Figure 5C). The Golgi stack cisternae and associated vesicles contained electron-dense substances, although not so electron-dense as the mature secretory vesicles (Figure 5D). On the other hand, in mucous cells the secretory vesicles were closely aggregated and many had fused forming vacuoles of diverse sizes filling most of the cytoplasm. These vesicles had an electron-dense spot and filaments embedded in a matrix of lower electron density (Figure 5E). In these cells the cisternae of rough endoplasmic reticulum were dilated and the Golgi stacks cisternae had an electron-lucent content (Figure 5F). In *A. fascicularis* both types of secretory cells contained electron-dense vesicles, the difference residing in vesicle diameter as observed with the light microscope (Table 1). An additional difference was found in the rough endoplasmic reticulum of these cells. The ones with the smaller secretory vesicles contained dilated cisternae, whereas in the cells with larger vesicles the cisternae were mostly flattened (Figure 6A,B). In both species, basal cells with round and oval electron-dense vesicles in a clear cytoplasm were frequently found in ultrathin sections of the oesophageal epithelium. However, in average the vesicles of basal cells were larger in *C. angulata* (Table 1). The cytoplasmic extensions of these cells were intertwined with the basal region of the absorptive cells, but a few basal cells with a long apical process were also observed (Figure 6C,D). However, it was not possible to see if these apical processes reached the lumen of the oesophagus. The outer layer of the oesophageal wall included smooth muscle cells and many nerves containing several axons and some glial cells, embedded in the connective tissue matrix (Figure 6D).

3.3. Histology and Ultrastructure of the Stomach

The stomach epithelium was identical in both species, being mainly formed by elongated absorptive cells with a central nucleus (Figure 7A,B). Epithelial height measured in semithin sections ranged mostly from about 40 μm at the thinner points between ridges to near 100 μm in thicker ridges, without significant differences detect between the two species. A layer of connective tissue, muscle fibres and some nerves formed the outer layer of the stomach wall usually with a thickness of 15–25 μm in *C. angulata* (Figure 7A) and 5–15 μm in *A. fascicularis*. With the light microscope only one type of secretory cells could be recognized in this epithelium, looking similar in both species. Their basophilic secretory vesicles stained dark blue with methylene blue and azure II, and were also strongly stained by PAS and tetrazonium reactions, but were not stained by Alcian blue (Figure 7B–F) (Table 1). Additionally, the microvilli layer covering the absorptive cells was stained by the PAS reaction in *A. fascicularis* (Figure 7E). As in the oesophagus, basal cells were present in the epithelium and their vesicles were stained by the tetrazonium coupling reaction for protein detection (Figure 7G).

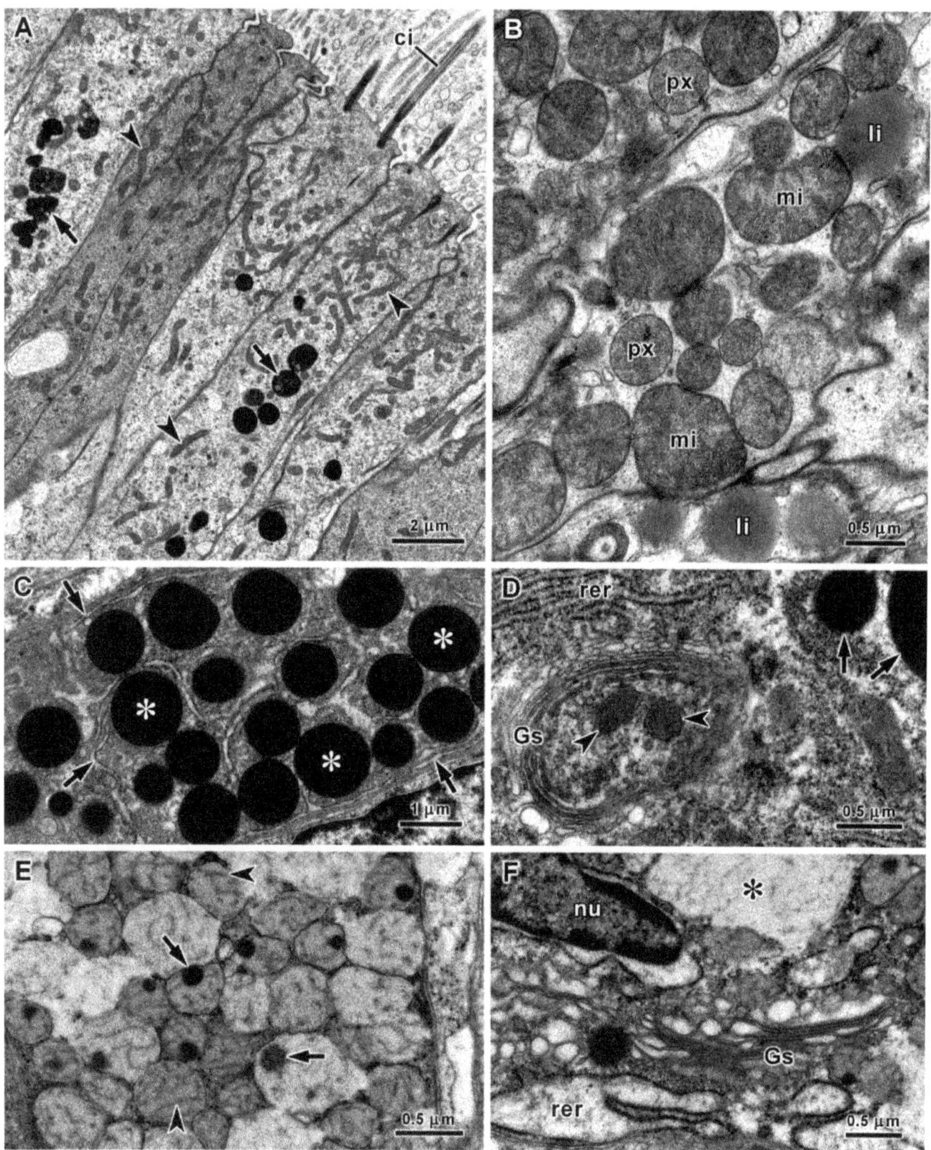

Figure 5. Ultrastructure of the oesophagus in *C. angulata* (**A,C–F**) and *A. fascicularis* (**B**). (**A**) Ciliated absorptive cells with several mitochondria (arrowheads) and lysosomes (arrows) in the apical cytoplasm. (**B**) Mitochondria (mi), peroxisomes (px) and lipid droplets (li) in the basal region of an absorptive cell. (**C**) Basophilic cell with electron-dense secretory vesicles (asterisks) and flattened cisternae of rough endoplasmic reticulum (arrows). (**D**) Golgi stack (Gs) and associated vesicles (arrowheads) with a content less electron-dense than the mature secretory vesicles (arrows) in a basophilic cell. (**E**) In mucus-secreting cells vesicles contained an electron-dense spot (arrows) and filaments (arrowheads). (**F**) Golgi stack (Gs) and dilated cisternae of rough endoplasmic reticulum (rer) in a mucus-secreting cell. Vesicle fusion created vacuoles with secretory products (asterisk). ci—cilium, nu—nucleus, rer—rough endoplasmic reticulum.

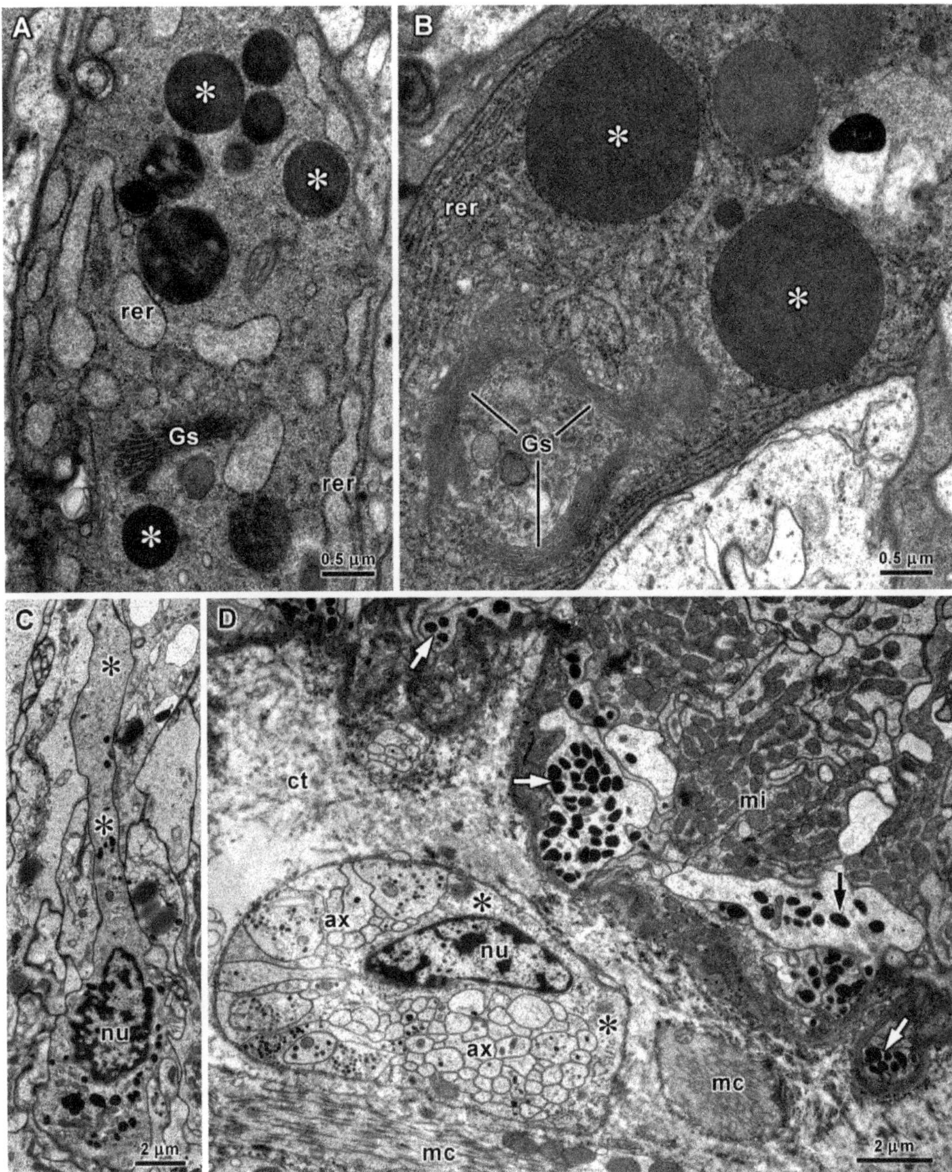

Figure 6. Ultrastructure of the oesophagus in *A. fascicularis*. (**A**) Basophilic cell with small electron-dense secretory vesicles (asterisks), dilated rough endoplasmic reticulum cisternae (rer) and a Golgi stack (Gs). (**B**) Basophilic cell with large electron-dense secretory vesicles (asterisks), flattened rough endoplasmic reticulum cisternae (rer) and Golgi stacks (Gs). (**C**) Basal cell with a long apical process (asterisks). (**D**) Basal region of the epithelium with basal cells containing electron-dense secretory vesicles (arrows). Many mitochondria (mi) are visible in the basal region of the absorptive cells. Muscle cells (mc) and a nerve section with several axons (ax) and a glial cell (asterisks) can be seen below the epithelium. ct—connective tissue matrix, nu—nuclei.

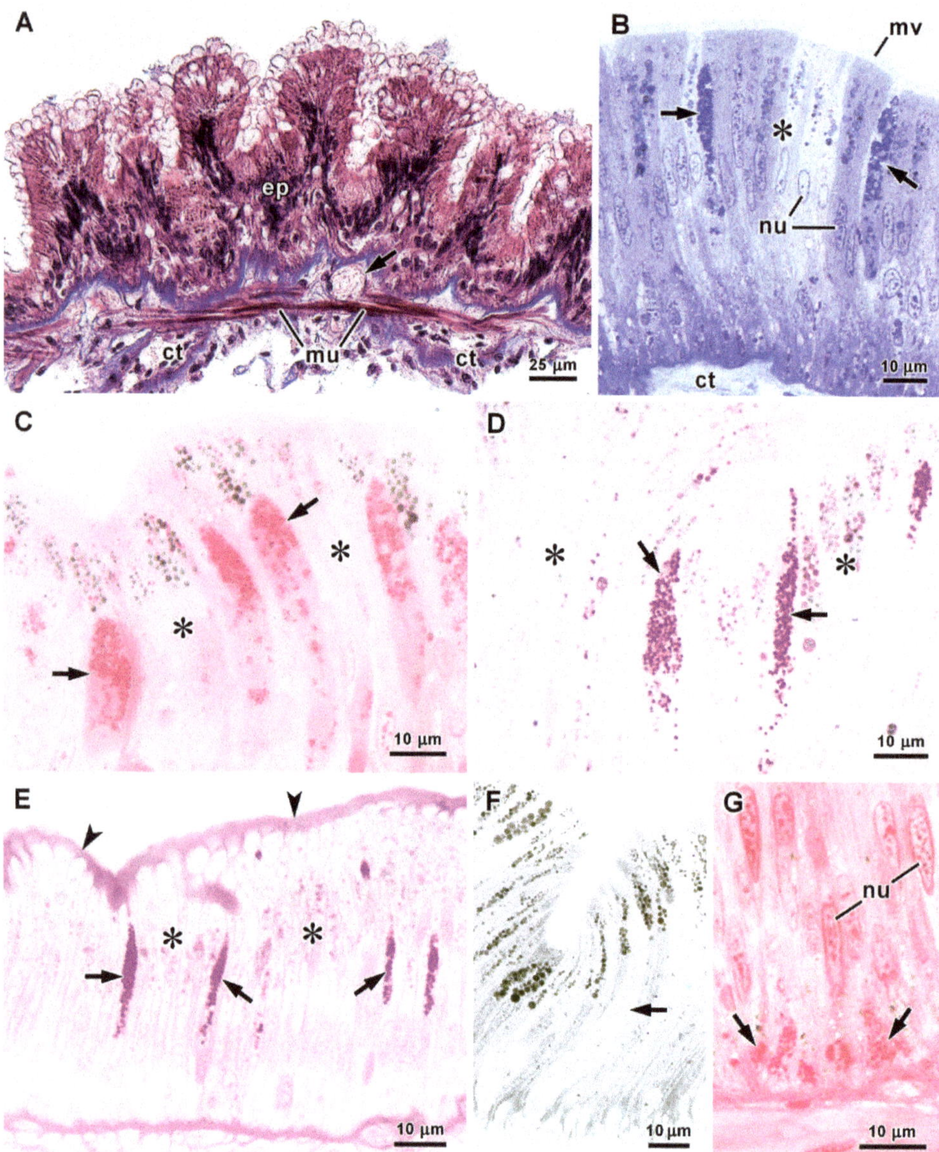

Figure 7. Stomach of *C. angulata* (**A–D,F,G**) and *A. fascicularis* (**E**) in histological (**A**) and semithin (**B–G**) sections. (**A**) Stomach wall stained by Masson's trichrome. A transverse section of a nerve (arrow) is visible below the epithelium (ep). (**B–D**) Absorptive cells (asterisks) and basophilic secretory cells (arrows) in the stomach of *C. angulata* stained by methylene blue and azure II (**B**), tetrazonium coupling reaction (**C**) and PAS reaction (**D**). (**E**) Secretory cells stained by PAS reaction (arrows) and absorptive cells (asterisks) in the stomach of *A. fascicularis*. The microvillous border is also stained (arrowheads). (**F**) Secretory cell (arrow) unstained by Alcian blue. (**G**) Basal cells (arrows) with vesicles stained by the tetrazonium coupling reaction. ct—connective tissue, mv—microvilli, mu—muscle fibres, nu—nuclei.

Transmission electron microscopy revealed the microvilli border of the stomach absorptive cells, but only some of them were ciliated. Vesicles, multivesicular bodies and many mitochondria were also observed in the apical region (Figure 8A). Lysosomes with variable sizes were abundant in the cytoplasm above the nucleus, and were stained with the electron-dense product of arylsulphatase detection (Figure 8B). Most lysosomes had diameters of 0.5–2.5 µm in ultrathin sections, but the largest ones that were found reached 5 µm. Some Golgi stacks and endoplasmic reticulum cisternae were also present. Glycogen deposits were abundant in most cells (Figure 8C), and especially in the basal region many peroxisomes were observed near mitochondria and lipid droplets. The peroxisomes of the stomach epithelium with 0.4–0.8 µm in diameter in ultrathin sections contained electron-dense inclusions in the matrix (Figure 8D). Deep cell membrane invaginations occurred in the basal region of the absorptive cells. The ultrastructural features of stomach secretory cells were identical in both species. Numerous electron-dense secretory vesicles occurred in the cytoplasm above the nucleus, with a tendency towards slightly larger vesicles in *A. fascicularis* (Table 1). Extensive Golgi stacks formed by cisternae with an electron-dense content and a great number of rough endoplasmic reticulum cisternae were other features of these cells (Figure 9A,B). Basal cells with oval electron-dense vesicles, identical to the basal cells of the oesophagus, were abundant in the stomach epithelium and in some ultrathin sections it was possible to observe close contact between nerves and basal cells (Figure 9C). A few intraepithelial neurons with a large round nucleus and many small electron-dense vesicles in the cytoplasm were also observed at the base of the stomach epithelium (Figure 9D).

3.4. Histology and Ultrastructure of the Intestine

Light microscopy observations of the intestine of both species revealed an epithelium with a wavy apical surface caused by differences in cell height. The intestinal epithelium ranged in height mostly between 40 µm and 100 µm and was largely formed by ciliated columnar absorptive cells (Figure 10A,B). The amount and intracellular distribution of lipid droplets in intestinal epithelial cells were variable, but numerous lipid droplets were frequently concentrated in the basal region of the absorptive cells (Figure 10C). Basophilic secretory cells containing small dark blue vesicles in semithin sections stained by methylene blue and azure II were present in the anterior intestine of both species (Figure 10D,E). Secretory cells in the anterior portion of *C. angulata* intestine were strongly stained by tetrazonium and PAS reactions (Figure 10F,G), but not by Alcian blue. On the other hand, in the anterior intestine of *A. fascicularis* secretory cells were stained by Alcian blue (Figure 10H) as well as by PAS and tetrazonium reactions (Table 1). In the valve between the anterior and posterior intestine of *A. fascicularis* the tall epithelial ridges further reduce the luminal space in this short narrow segment of the intestine (Figure 11A). Mucus-secreting cells with a basal nucleus were very abundant in the valve epithelium, alternating with very thin ciliated cells with a more apically located nucleus. This difference in the position of nuclei was more evident in thicker zones of the epithelium, in which two levels of nuclei were clearly visible (Figure 11A,B). In these mucous cells individualised secretory vesicles could not be recognized by light microscopy. Instead, a large mass of secretion moderately stained by methylene blue and azure II filled the cytoplasm. Most were strongly stained by PAS reaction (Figure 11C), but staining of histological sections with both PAS reaction and Alcian blue revealed heterogeneity among these mucous cells. Some were stained only by PAS reaction, others mainly by Alcian blue and still others showed double staining (Figure 11D). Secretion in these cells was weakly stained by the tetrazonium coupling reaction indicating a low protein content (Figure 11E). Secretory cells were rarely found in the posterior intestine of *A. fascicularis*, and were of the basophilic type as in the anterior intestine of this species. In *C. angulata* a very large number mucus-secreting cells strongly stained by Alcian blue were present in a short segment of the intestine that could correspond to the valve of *A. fascicularis* intestine. In the posterior intestine of *C. angulata* mucous cells were found in low numbers. These cells had a basal nucleus and contained numerous vesicles unstained by methylene blue and azure II in semithin sections (Figure 11F). These secretory vesicles were strongly stained by Alcian blue (Figure 11G), but not at all by tetrazonium (Figure 11H) or PAS reactions (Table 1). Additionally, some basophilic cells were found in the posterior intestine of *C. angulata.* Basal cells were also observed in the intestinal epithelium of both species (Figure 10D). The outer layer of the intestine wall consisting of connective tissue, muscle cells and some nerves in general had a thickness of 15–25 µm, but only 5–15 µm in the posterior intestine of *A. fascicularis.*

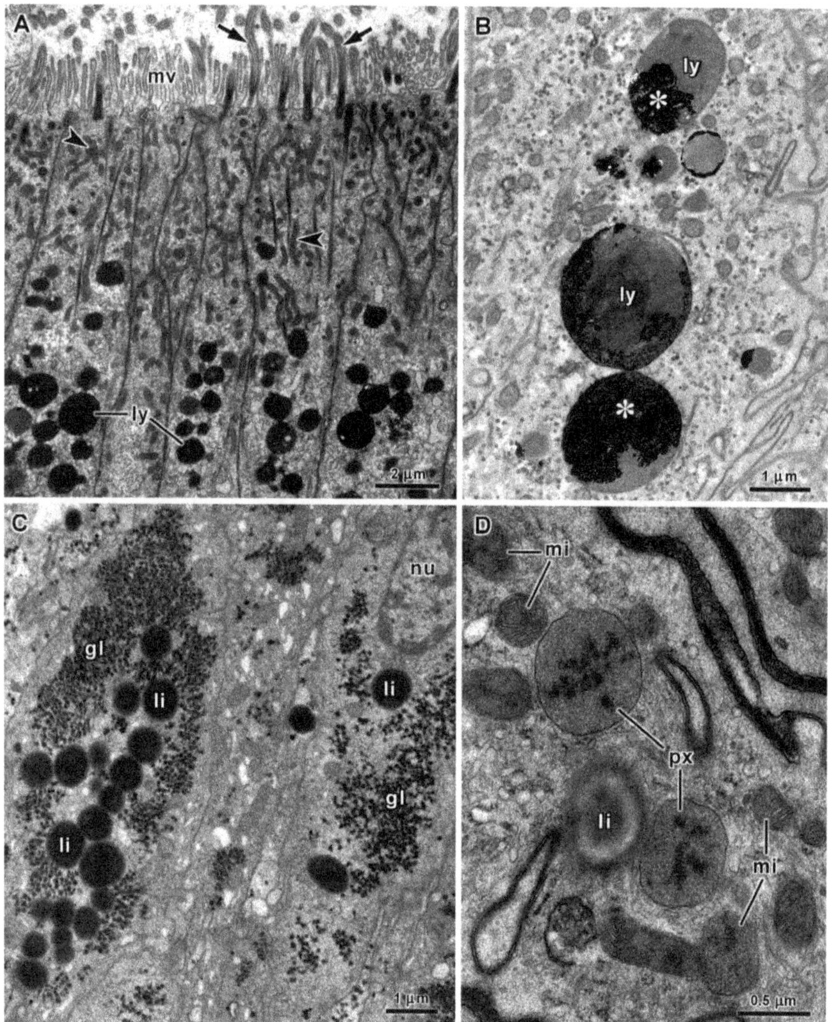

Figure 8. Ultrastructure of stomach absorptive cells in *C. angulata* (**A,B**) and *A. fascicularis* (**C,D**). (**A**) Apical region of the epithelium with a border of cilia (arrows) and microvilli (mv). Lysosomes (ly) and mitochondria (arrowheads) are abundant in the cytoplasm above the nucleus. (**B**) Lysosomes (ly) with electron-dense deposits (asterisks) resulting from arylsulphatase detection. (**C**) Glycogen deposits (gl) and lipid droplets (li) in the cytoplasm below the nucleus (nu). Ultrathin section stained with tannic acid and uranyl acetate. (**D**) Peroxisomes (px) in association with mitochondria (mi) and a lipid droplet (li).

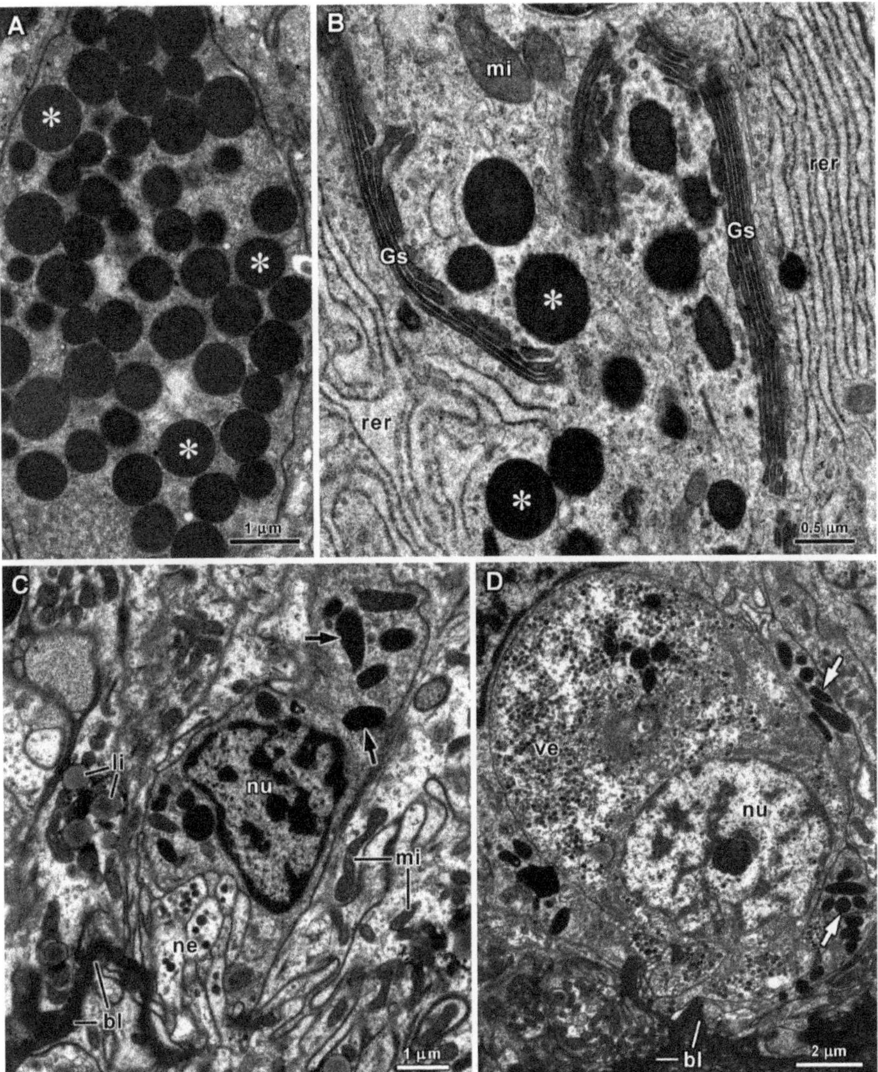

Figure 9. Ultrastructure of the stomach in *A. fascicularis* (**A**) and *C. angulata* (**B–D**). (**A**) Electron-dense vesicles (asterisks) in the apical region of a basophilic secretory cell. (**B**) Golgi stacks (Gs) and rough endoplasmic reticulum cisternae (rer) in a basophilic secretory cell containing electron-dense vesicles (asterisks). (**C**) Basal cell with electron-dense vesicles (arrows) in connection with a nerve terminal (ne). (**D**) Intraepithelial neuron with a large number of small vesicles (ve) in the cytoplasm. Basal cells with the electron-dense secretory vesicles (arrows) are visible around the neuron. bl—basal lamina, li—lipid droplets, mi—mitochondria, nu—nuclei.

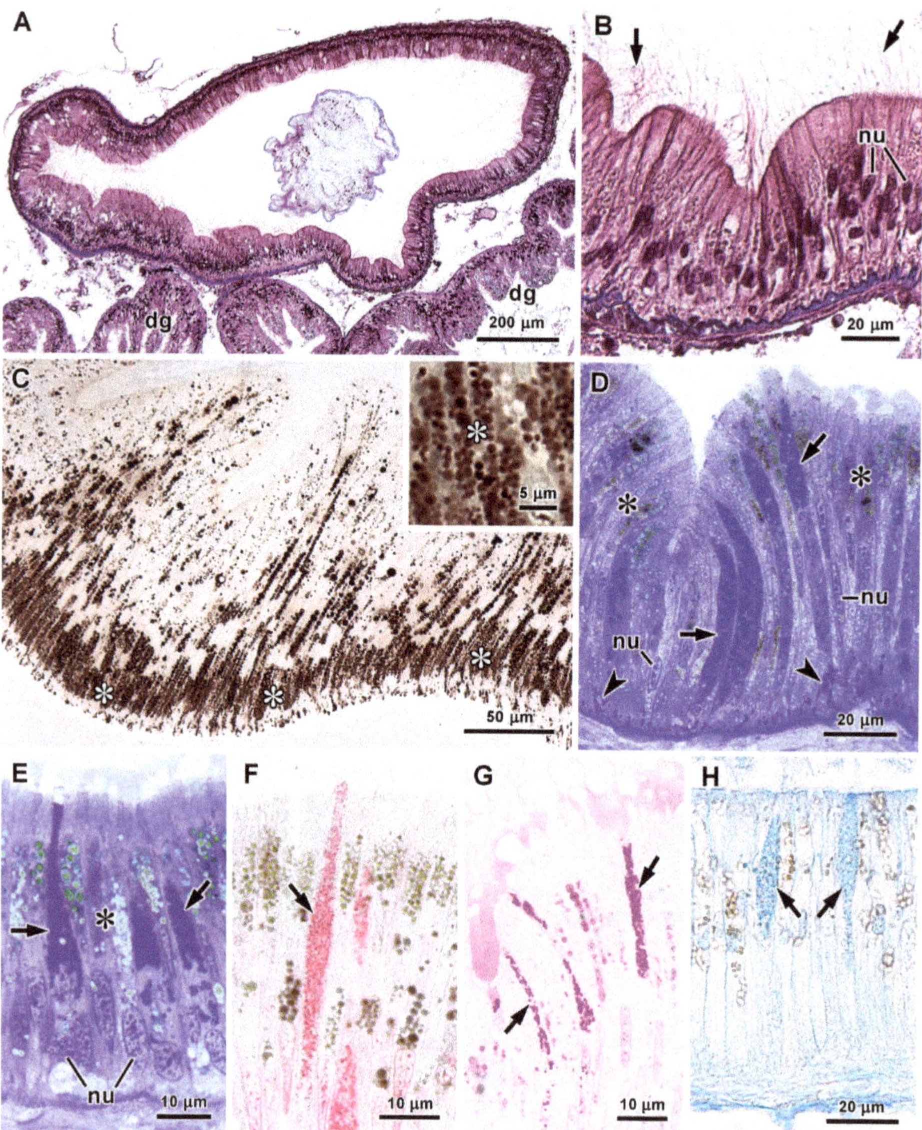

Figure 10. Intestine *C. angulata* (**A–D,F,G**) and *A. fascicularis* (**E,H**) in histological (**A,B,H**) and semithin (**C–G**) sections. (**A,B**) Transverse sections of the intestine stained by Masson's trichrome. The epithelium is covered by cilia (arrows in (**B**)). ((**C**), inset) Lipid droplets stained by Sudan black in the posterior intestine. Most lipid droplets (asterisks) are concentrated in the basal region of the absorptive cells. (**D**) Basophilic secretory cells (arrows), absorptive cells (asterisks) and basal cells (arrowheads) in the anterior intestine of *C. angulata*. Methylene blue and azure II staining. (**E**) Basophilic secretory cells (arrows) and absorptive cells (asterisk) in the anterior intestine of *A. fascicularis*. Methylene blue and azure II staining. (**F,G**) Secretory cells (arrows) stained by tetrazonium reaction (**F**) and PAS reaction (**G**) in the anterior intestine of *C. angulata*. (**H**) Secretory cells (arrows) stained by Alcian blue in the anterior intestine of *A. fascicularis*. dg—digestive gland, nu—nuclei.

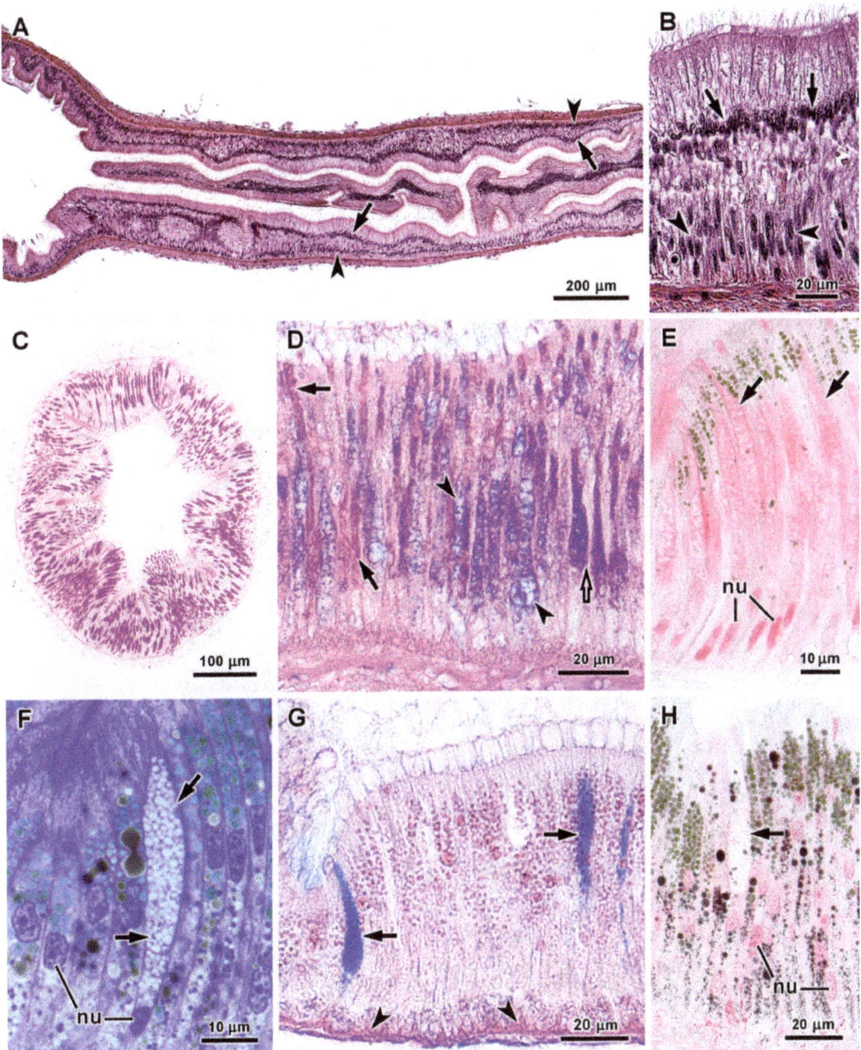

Figure 11. Intestine of *A. fascicularis* (**A–E**) and *C. angulata* (**F–H**) in histological (**A,B,D,G**) and semithin (**C,E,F,H**) sections. (**A,B**) Longitudinal (parasagittal) section of the valve between the anterior and posterior intestine of *A. fascicularis* stained by Masson's trichrome. The basal row of nuclei correspond to the nuclei of mucus-secreting cells (arrowheads) and the nuclei in the upper row belong to absorptive ciliated cells (arrows). (**C**) Transverse section of the valve with the mucus-secreting cells strongly stained by PAS reaction. (**D**) In the valve epithelium some mucus-secreting cell were stained only by PAS reaction (arrows), others mostly by Alcian blue (arrowheads) and still other were stained by both (open arrow). (**E**) Mucus-secreting cells of the valve (arrows) weakly stained by the tetrazonium coupling reaction. (**F–H**) Mucus-secreting cells (arrows) in the posterior intestine of *C. angulata*, unstained by methylene blue and azure II (**F**), strongly stained by Alcian blue (**G**) and unstained by the tetrazonium coupling reaction (**H**). The basal lamina is stained by PAS reaction (arrowheads in **G**). nu—nuclei.

As in other parts of the digestive tract, the absorptive cells of the intestine presented ultrastructural evidence of endocytic activity and intracellular digestion, namely, cell membrane pits, multivesicular bodies and lysosomes (Figure 12A–C). In *A. fascicularis* short cisternae with electron-dense content were present in the cytoplasm beneath the cell membrane, which were probably also related with endocytosis because a similar electron-dense content was found in multivesicular bodies (Figure 12A,B). In intestinal absorptive cells of both species mitochondria were abundant in the apical region in association with some peroxisomes. Lysosomes with electron-dense content were prominent in the cytoplasm above the nucleus and most had a diameter around 0.5–2.0 μm (Figure 12C), with the largest ones that were observed attaining about 3 μm in diameter. The detection of arylsulphatase activity in these organelles confirmed that they are active lysosomes. Deposits of glycogen granules were also present. Mainly in the posterior intestine, several peroxisomes were observed in association with large amounts of lipid droplets and mitochondria in the cytoplasm below the nucleus. These peroxisomes presented a granular matrix without cores (Figure 12D), and most had diameters of 0.5–1.5 μm in ultrathin sections.

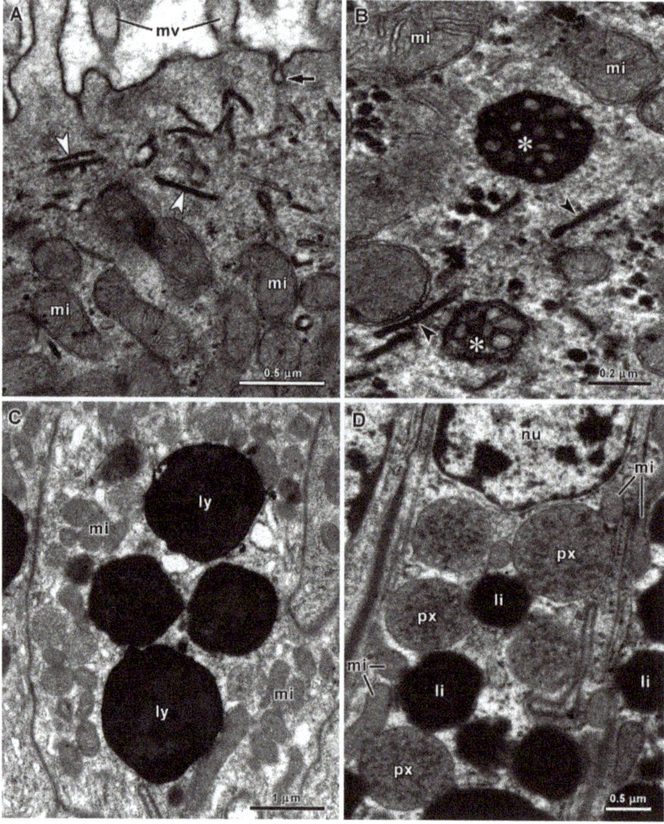

Figure 12. Ultrastructure of intestinal absorptive cells in *A. fascicularis* (**A,B,D**) and *C. angulata* (**C**). (**A,B**) A cell membrane pit (arrow), several mitochondria (mi), cisternae with electron-dense content (arrowheads) and multivesicular bodies (asterisks) in the apical region of absorptive cells. (**C**) Electron-dense lysosomes (ly) in the cytoplasm above de nucleus. (**D**) Several large peroxisomes (px) associated with lipid droplets (li) and mitochondria (mi) in the cytoplasm below the nucleus (nu). mi—mitochondria, mv—microvilli, nu—nucleus.

In both species, the secretory cells of the anterior intestine contained a large number of electron-dense vesicles (Figure 13A) with a tendency towards larger diameters in *A. fascicularis* (Table 1). These cells were rich in rough endoplasmic reticulum cisternae and the cisternae of their Golgi stacks had an electron-dense content (Figure 13B). In the valve between the anterior and posterior intestine of *A. fascicularis*, the mucus-secreting cells contained an accumulation of partially fused vesicles with flocculent material in an electron-lucent background, dilated rough endoplasmic reticulum cisternae and long Golgi stacks (Figure 13C,D). In the posterior intestine of *C. angulata* the cytoplasm of the mucus-secreting cells was almost entirely filled by round vesicles with low or median electron density. A few rough endoplasmic reticulum cisternae were present between the secretory vesicles (Figigue 14A). Some secretory cells with electron-dense secretory vesicles were also found in the posterior intestine of *A. fascicularis* (Figure 14B). However, in these secretory cells dome-shaped mitochondria (ring-shaped in some ultrathin sections) were frequently seen surrounding dilated rough endoplasmic reticulum cisternae (Figure 14C). This peculiar association between mitochondria and rough endoplasmic reticulum cisternae was not observed in any other cells of both species. Basal cells of the intestinal epithelium exhibited the same ultrastructural features and association with nerves observed in the oesophagus and stomach (Figure 14D).

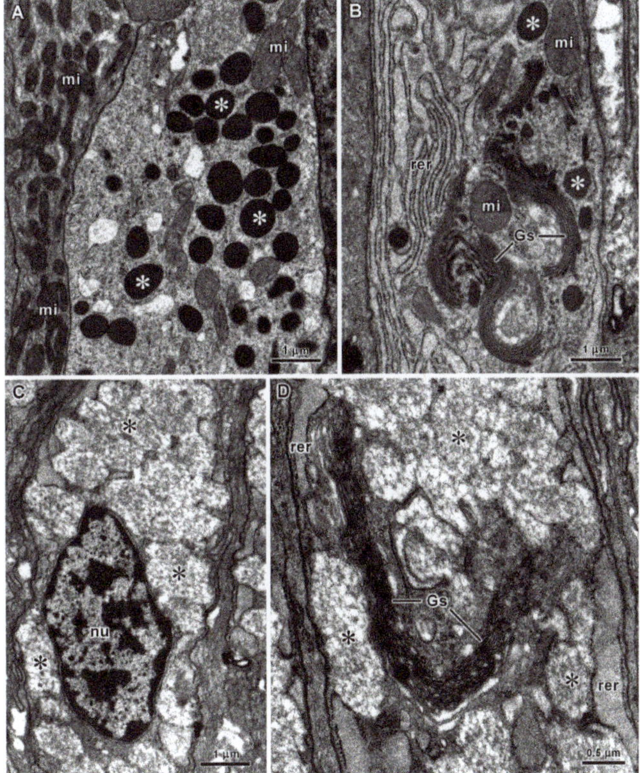

Figure 13. Ultrastructure of intestinal secretory cells in *C. angulata* (**A,B**) and *A. fascicularis* (**C,D**). (**A,B**) Electron-dense vesicles (asterisks), rough endoplasmic reticulum cisternae (rer) and Golgi stacks (Gs) in basophilic secretory cells of the anterior intestine. (**C,D**) Mucous cells of the valve between the anterior and posterior intestine in *A. fascicularis* with partially fused secretory vesicles (asterisks). These cells contain dilated rough endoplasmic reticulum cisternae (rer) and long Golgi stacks (Gs). nu—nucleus.

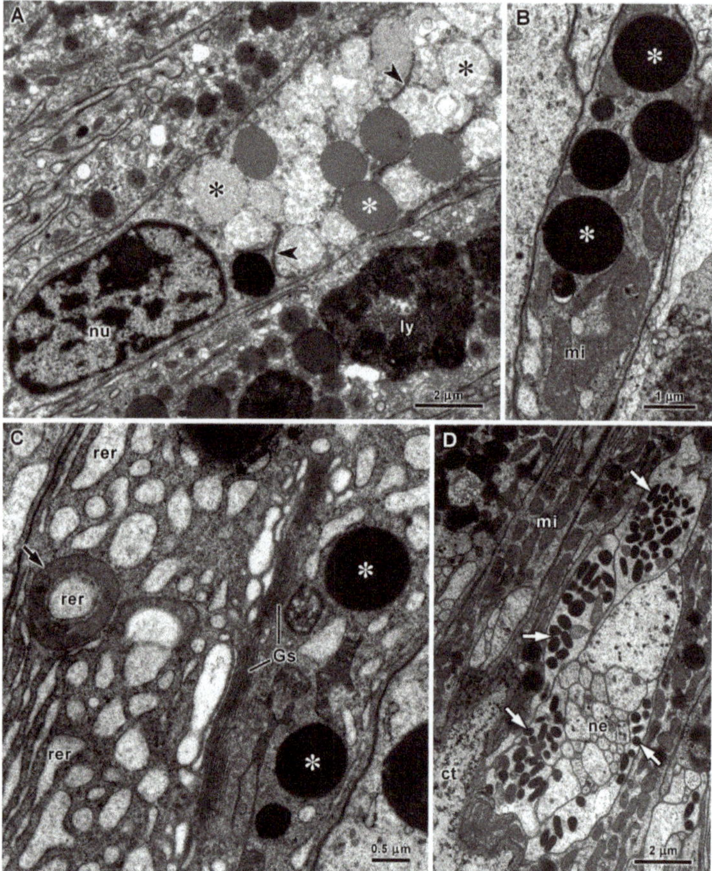

Figure 14. Ultrastructure of the posterior intestine in *C. angulata* (**A**) and *A. fascicularis* (**B–D**). (**A**) Mucus-secreting cell with vesicles of low and median electron density (asterisks) and a few rough endoplasmic reticulum cisternae (arrowheads). (**B**) Apical region of a narrow basophilic cell with electron-dense secretory vesicles (asterisks). (**C**) Electron-dense vesicles (asterisks), rough endoplasmic reticulum cisternae (rer), Golgi stacks (Gs) and a mitochondrion with a ring-shaped section (arrow) surrounding a portion of endoplasmic reticulum in a basophilic secretory cell. (**D**) Basal cells with oval electron-dense vesicles (arrows) around an intraepithelial nerve terminal (ne). ct—connective tissue, ly—lysosome, mi—mitochondria, nu—nucleus.

4. Discussion

This article presents the results of the first histochemical and ultrastructural investigation of the oesophagus, stomach and intestine of chitons. The species *Chaetopleura angulata* and *Acanthochitona fascicularis* were chosen for this propose because they were previously employed for the study of the oesophageal pouches [9]. Both species occur intertidally on the Portuguese coast and are common at the collection site. They belong, respectively, to the families Chaetopleuridae (suborder Chitonina) and Acanthochitonidae (suborder Acanthochitonina) of the order Chitonida that comprises around 80% of the living species of chitons [23]. The stomach content of the two species revealed a diversified omnivorous diet, supporting the growing evidence that in general intertidal chitons are omnivorous [11,12]. Previous reports indicated the consumption of barnacles by *C. angulata* and based on that information the species was classified as carnivorous [24], but the present results clearly show that it is omnivorous. The ingestion of small crustaceans by *C. angulata* was confirmed, but remains of invertebrates were not found in the digestive tract of *A. fascicularis*. Although substantial amounts of unidentifiable residues were present, algae largely predominated in the stomach content of the *A. fascicularis* specimens that were analysed. Additionally, the intestine-body length ratios of the species included in this study (2.4–3.0) fit well within the ratios found for other omnivorous chitons [8].

The digestive process is still much less studied in polyplacophorans than in gastropods, bivalves or cephalopods [25]. Nevertheless, it is recognized that digestion in polyplacophorans follows the general pattern known for other molluscs, starting in the lumen of the digestive tract with secreted enzymes and ending intracellularly in lysosomes. As in other chitons [5] and generally in molluscs [26], the digestive tract epithelium in *C. angulata* and *A. fascicularis* consists of ciliated and non-ciliated columnar absorptive cells and secretory cells. The observation of cell membrane pits, vesicles and multivesicular bodies, which belong to the endolysosomal system [27], are indicative of endocytosis in the apical region of absorptive cells as experimentally demonstrated in other molluscs [28–30]. By this process, ultrafine nutritive particles resulting from extracellular digestion in the digestive tract lumen are introduce in the absorptive cells to be digested in lysosomes [31], which are abundant organelles in the entire digestive tract epithelium of these two species of chitons, as well as in their oesophageal pouches [9]. Additionally, small molecules such as simple sugars and amino acids can be absorbed directly through the cell membrane of the microvilli border of the digestive tract epithelium, as demonstrated in other molluscs [28,29,32]. Although in molluscs the digestive gland is considered the main site of nutrient absorption, intracellular digestion and storage of reserves, these functions are shared with epithelial cells of the digestive tract [26]. Mitochondria were particularly abundant in the apical and basal regions of absorptive cells. In the apical region, these organelles must be required to supply energy for ciliary motion and endocytosis. The concentration of mitochondria in the basal region of the digestive tract epithelium also denotes a need for high ATP production in this region, naturally required for active transport of metabolites and ions across the basal cell membrane. Moreover, the required energy can be obtained from glycogen and lipid reserves that are present in absorptive cells. The association of lipid droplets with peroxisomes mainly at the basal region of absorptive cells suggests the use of fatty acids to produce acetyl-CoA through peroxisomal β-oxidation, in order to fuel the citric acid cycle in nearby mitochondria and ultimately produce ATP [33]. Data are not available for polyplacophorans, but fatty acid β-oxidation is a standard metabolic pathway of peroxisomes [34]. Enzymes of this pathway were detected in mussel peroxisomes [35,36] and should also be present in digestive system of other molluscs. Although not always present, electron-dense crystalline cores are common in peroxisomes of molluscs and other taxa [37–39], and were found in peroxisomes of absorptive cells in *A. fascicularis* oesophageal pouches [9] and in peroxisomes of chiton digestive gland [7]. However, cores were not seen in digestive tract peroxisomes of *A. fascicularis* and *C. angulata*, in which only some electron-dense spots were observed in the peroxisomal matrix. In molluscs, peroxisomes are usually larger and more abundant in the digestive gland [7,37–39]. Peroxisomes not exciding 0.5 μm in diameter closely associated with mitochondria and lipid droplets were reported in digestive tract epithelial cell of gastropods [40–43]. However, in the intestine of *A. fascicularis* and *C. angulata* peroxisomes are larger and numerous, denoting a higher importance of peroxisomal metabolism in these cells where lipid reserves are often abundant.

Though the absorptive cells of the digestive tract epithelium were found to be similar in these two species of chitons, some differences were detected in their secretory cells. Secretory cells in the digestive tract of *A. fascicularis* and *C. angulata* can be divided in two main categories: basophilic and mucous. Cells containing secretory vesicles with high electron density when observed by transmission electron microscopy and displaying a dark blue colour in semithin sections stained by methylene blue and azure II were designated as basophilic cells. This designation is applied to the secretory cells of the digestive gland responsible for secretion of enzymes for extracellular digestion in molluscs [26], and was also used for one type of secretory cells of the glandular oesophageal pouches of chitons that could as well be responsible for secretion of digestive enzymes [9]. These cells containing electron-dense secretory vesicles with a high protein concentration are rich in rough endoplasmic reticulum cisternae and could be considered serous cells [44]. However, the histochemical methods indicate that the secretory vesicles of chiton basophilic cells are also rich in polysaccharides, and in some cases including acidic polysaccharides that are more typical of mucus-secreting cells. Secretory cells with serous features were also found in the digestive tract of other molluscs, and although precise knowledge about their functions are still lacking they could be responsible for secretion of some digestive enzymes [45–48]. In alternative, it was also suggested that intestinal secretory cells could produce a proteinaceous cement to coat the faecal rods with a protective layer to prevent their disintegration [32]. Mucous cells typically contain secretory vesicles with low electron-density rich in acid polysaccharides [49]. Mucous cells are abundant in the digestive tract of several herbivorous gastropods [43,45,48,50,51] and in the oesophageal pouches of *A. fascicularis* and *C. angulata* [9]. However, cells of this type were not common in the digestive tract of these two chiton species, except in the intestinal valve where mucous cells were very abundant. Mucous secretion can protect the digestive tract epithelium and agglutinate particles. It is also known that mucin MUC2 secreted by vertebrate intestinal goblet cells is a highly glycosylated glycoprotein with immunoregulatory functions [52]. Thus, secretion of glycoproteins with defensive and regulatory activities is another possible function for the intestinal secretory cells of chitons and other molluscs. Therefore, without further knowledge about the precise physiological role of the different secretory cell types found in the digestive tract of chitons and other molluscs it is not possible to evaluate the full meaning of the histological differences concerning these cells.

The basal cells that were observed throughout the digestive tract epithelium of *C. angulata* and *A. fascicularis* presented histological and ultrastructural features of enteroendocrine cells, namely, clear cytoplasm and basally located electron-dense secretory vesicles (also called secretory granules) [53,54]. Morphologically identical basal cells were recently reported in the oesophageal pouches of these two chitons species [9]. However, these cells that are detectable in semithin sections were not mentioned in earlier histological studies of polyplacophoran digestive system [5], probably because they are not evident in classical histological sections. Enteroendocrine cells occur along the digestive tract of both vertebrates and invertebrates, and produce peptidic hormones that are accumulated in secretory vesicles ready to be released at the base of the epithelium in response to specific stimuli [55,56]. Hence, the histochemical detection of proteic substances in the vesicles of chiton basal cells by the tetrazonium coupling reaction is compatible with the production and storage of peptidic hormones in these cells. The presence of endocrine cells in the digestive tract epithelium of gastropods and bivalves was confirmed by immunostaining with antibodies against vertebrate peptidic hormones such as insulin, gastrin, somatostatin and glucagon [46,57–59]. More recently, insulin related peptides were identified and sequenced in some molluscs and other invertebrates [60]. Therefore, similar hormones should also be produced by the enteroendocrine cells of polyplacophorans for metabolic regulation. In the digestive tract of gastropods and bivalves enteroendocrine cells are isolated and scattered in the epithelium [43,46,57,59,61]. In mammals these cells are also rare and dispersed, corresponding to less than 1% of the total epithelial cells of the intestine [56,62]. Conversely, the enteroendocrine-like basal cells are numerous in the entire digestive tract of chitons, forming a distinct layer at the base of the epithelium. Since contact of these cells with the digestive tract lumen was not observed, at least most of them must be enteroendocrine cells of the closed type. Enteroendocrine cells of the open type reach the digestive tract lumen and can release hormones in direct response to the molecules present in the lumen, whereas enteroendocrine cells of close type are believed to be indirectly activated by those chemical stimuli [55]. In the intestine of the freshwater snail *Planorbarius corneus* and the sea hare *Aplysia depilans* the enteroendocrine cells also appear to be isolated from the lumen [43,46].

However, in the land snail *Helix aspersa* and in freshwater bivalves the apical process of the enteroendocrine cells reached the lumen of the intestine [57,61]. In the intestine of the gastropod *Vivaparus ater* both close and open types of enteroendocrine cells were reported [59]. Another important aspect is the connection between the enteroendocrine cells and nerves, which was frequently observed along the digestive tract of *C. angulata* and *A. fascicularis*, as in other molluscs [43,46,61]. Due to these connections, enteroendocrine cells are a fundamental component of a neuroepithelial circuit that provides an interaction between the digestive tract and the nervous system [63].

The main histological differences between the digestive tract of *C. angulata* and *A. fascicularis* concern the secretory cells. These histological differences could be related to diet or being due to the taxonomic distance between these species that belong to different suborders. Although both species included in this study consume foraminiferans, unicellular and multicellular algae, remains of crustaceans were found only in the stomach of *C. angulata*. However, without data on other chiton species with similar and distinct diets it is premature to establish a correlation between feeding habits and digestive system histology.

Supplementary Materials: The following supporting information was downloaded at: https://www.mdpi.com/article/10.3390/jmse10020160/s1, Figure S1: Species and collection site. (A,B) Location of the collection site (arrows). (C,D) Specimens of *Chaetopleura angulata* (C) and *Acanthochitona fascicularis* (D) on roof tiles that were dumped in large amounts at the collection site and are now colonized by intertidal species.

Author Contributions: A.L.-d.-C. research and writing; G.C., Â.A. and E.O. research. All authors have read and agreed to the published version of the manuscript.

Funding: The study was supported by funding from the Institute of Biomedical Sciences Abel Salazar (ICBAS) of the University of Porto (Portugal), and FCT—Fundação para a Ciência e a Tecnologia (Portugal), through the strategic project UIDB/04292/2020 granted to MARE.

Institutional Review Board Statement: Not required in this case.

Informed Consent Statement: Not applicable.

Data Availability Statement: Microscopy slides kept in the Department of Microscopy (ICBAS-UP, Porto, Portugal) can be consulted on request.

Acknowledgments: The authors thank Paula Teixeira and Aurora Rodrigues for technical assistance.

Conflicts of Interest: The authors declare no conflict of interest.

References

1. Eernisse, D.J.; Reynolds, P.D. Polyplacophora. In *Microscopic Anatomy of Invertebrates*; Harrison, F.W., Kohn, A.J., Mollusca, I., Eds.; Wiley-Liss: New York, NY, USA, 1994; Volume 5, pp. 55–110.
2. Todt, C.; Okusu, A.; Schander, C.; Schwabe, E. Solenogastres, Caudofoveata, and Polyplacophora. In *Phylogeny and Evolution of the Mollusca*; Ponder, W.F., Lindberg, D.R., Eds.; University of California Press: Berkeley, CA, USA, 2008; pp. 71–96.
3. Sigwart, J.D.; Vermeij, G.J.; Hoyer, P. Why do chitons curl into a ball? *Biol. Lett.* **2019**, *15*, 20190429. [CrossRef] [PubMed]
4. Ponder, W.F.; Lindberg, D.R.; Ponder, J.M. *Biology and Evolution of the Mollusca*; CRC Press: Boca Raton, FL, USA, 2020; Volume 2.
5. Fretter, V. The structure and function of the alimentary canal of some species of Polyplacophora (Mollusca). *Trans. R. Soc. Edin.* **1937**, *49*, 119–164. [CrossRef]
6. Greenfield, M.L. Feeding and gut physiology in *Acanthopleura spinigera* (Mollusca). *J. Zool. Lond.* **1972**, *166*, 37–47. [CrossRef]
7. Lobo-da-Cunha, A. The peroxisomes of the hepatopancreas in two species of chitons. *Cell Tissue Res.* **1997**, *290*, 655–664. [CrossRef]
8. Sigwart, J.D.; Schwabe, E. Anatomy of the many feeding types in polyplacophoran molluscs. *Invert. Zool.* **2017**, *14*, 205–216. [CrossRef]
9. Lobo-da-Cunha, A.; Alves, A.; Oliveira, E.; Calado, G. Functional morphology of the glandular esophageal pouches of chitons (Mollusca, Polyplacophora). *J. Morphol.* **2021**, *282*, 355–367. [CrossRef]
10. Latyshev, N.A.; Khardin, A.S.; Kasyanov, S.P.; Ivanova, M.B. A study on the feeding ecology of chitons using analysis of gut contents and fatty acid markers. *J. Moll. Stud.* **2004**, *70*, 225–230. [CrossRef]
11. Camus, P.; Daroch, K.; Opazo, L. Potential for omnivory and apparent intraguild predation in rocky intertidal herbivore assemblages from northern Chile. *Mar. Ecol. Progr. Ser.* **2008**, *361*, 35–45. [CrossRef]

12. Camus, P.; Arancibia, P.; Ávila-Thieme, M. A trophic characterization of intertidal consumers on Chilean rocky shores. *Rev. Biol. Mar. Oceanogr.* **2013**, *48*, 431–450. [CrossRef]

13. Saito, H.; Okutani, T. Carnivorous habits of two species of the genus *Craspedochiton* (Polyplacophora: Acanthochitonidae). *J. Malacol. Soc. Aust.* **1992**, *13*, 55–63. [CrossRef]

14. Duperron, S.; Pottier, M.-A.; Léger, N.; Gaudron, S.M.; Puillandre, N.; Le Prieur, S.; Sigwart, J.D.; Ravaux, J.; Zbinden, M. A tale of two chitons: Is habitat specialisation linked to distinct associated bacterial communities? *FEMS Microbiol. Ecol.* **2013**, *83*, 552–567. [CrossRef] [PubMed]

15. Brooker, L.R.; Shaw, J. The chiton radula: A unique model for biomineralization studies. In *Advanced Topics in Biomineralization*; Seto, J., Ed.; InTech: Rijeka, Croatia, 2012; pp. 65–84. [CrossRef]

16. Boyle, P.R. Fine structure of the subradular organ of Lepidochitona cinereus (L), (Mollusca, Polyplacophora). *Cell Tissue Res.* **1975**, *162*, 411–417. [CrossRef] [PubMed]

17. Meeuse, B.J.D.; Fluegel, W. Carbohydrate-digesting enzymes in the sugar gland juice of Cryptochiton stelleri. *Arch. Neerl. Zool.* **1958**, *13*, 301–313. [CrossRef]

18. Saito, H.; Okutani, T. Taxonomy of Japanese species of the genera *Mopalia* and *Plaxiphora* (Polyplacophora: Mopaliidae). *Veliger* **1991**, *34*, 172–194.

19. Sannes, P.L.; Katsuyama, T.; Spicer, S. Tannic acid-metal salt sequences for light and electron microscopic localization of complex carbohydrates. *J. Histochem. Cytochem.* **1978**, *26*, 55–61. [CrossRef]

20. Ganter, P.; Jollès, G. *Histochimie Normal et Pathologique*; Gauthier-Villars: Paris, France, 1970.

21. Lane, B.P.; Europa, D.L. Differential staining of ultrathin sections of Epon-embedded tissues for light microscopy. *J. Histochem. Cytochem.* **1965**, *13*, 579–582. [CrossRef]

22. Hopsu-Havu, V.K.; Arstila, A.V.; Helminen, H.J.; Kalimo, H.O. Improvements in method for electron microscopic localization of arylsulphatase activity. *Histochemie* **1967**, *8*, 54–64. [CrossRef]

23. Sigwart, J.D.; Stoeger, I.; Knebelsberger, T.; Schwabe, E. Chiton phylogeny (Mollusca: Polyplacophora) and the placement of the enigmatic species *Choriplax grayi* (H. Adams & Angas). *Invertebr. Syst.* **2013**, *27*, 603–621. [CrossRef]

24. Kaas, P.; Van Belle, R.A. Monograph of living chitons (Mollusca: Polyplacophora). In *Suborder Ischnochitonina, Ischnochitonidae: Chaetopleurinae & Ischnochitoninae (pars)*; E. J. Brill and W. Backhuys: Leiden, The Netherlands, 1987; Volume 3.

25. Ponder, W.F.; Lindberg, D.R.; Ponder, J.M. *Biology and Evolution of the Mollusca*; CRC Press: Boca Raton, FL, USA, 2020; Volume 1.

26. Lobo-da-Cunha, A. Structure and function of the digestive system in molluscs. *Cell Tissue Res.* **2019**, *377*, 475–503. [CrossRef]

27. Klumperman, J.; Raposo, G. The complex ultrastructure of the endolysosomal system. *Cold Spring Harb. Perspect. Biol.* **2014**, *6*, a016857. [CrossRef]

28. Walker, G. The digestive system of the slug, *Agriolimax reticulatus* (Müller): Experiments on phagocytosis and nutrient absorption. *Proc. Malacol. Soc. Lond.* **1972**, *40*, 33–43. [CrossRef]

29. Bush, M.S. The ultrastructure and function of the oesophagus of *Patella vulgata*. *J. Moll. Stud.* **1989**, *55*, 111–124. [CrossRef]

30. Bourne, N.B.; Jones, W.; Bowen, I.D. Endocytosis in the crop of the slug, Deroceras reticulatum (Müller) and the effects of the ingested molluscicides, metaldehyde and methiocarb. *J. Moll. Stud.* **1991**, *57*, 71–80. [CrossRef]

31. Huotari, J.; Helenius, A. Endosome maturation. *EMBO J.* **2011**, *30*, 3481–3500. [CrossRef] [PubMed]

32. Bush, M.S. The ultrastructure and function of the intestine of *Patella vulgata*. *J. Zool.* **1988**, *215*, 685–702. [CrossRef]

33. Schrader, M.; Kamoshita, M.; Islinger, M. Organelle interplay - peroxisome interactions in health and disease. *J. Inherit. Metab. Dis.* **2020**, *43*, 71–89. [CrossRef] [PubMed]

34. Islinger, M.; Cardoso, M.J.R.; Schrader, M. Be different—The diversity of peroxisomes in the animal kingdom. *Biochim. Biophys. Acta Mol. Cell Res.* **2010**, *1803*, 881–897. [CrossRef]

35. Cajaraville, M.P.; Völkl, A.; Fahimi, H.D. Peroxisomes in the digestive gland cells of the mussel *Mytilus galloprovincialis* Lmk. Biochemical, ultrastructural and immunocytochemical characterization. *Europ. J. Cell Biol.* **1992**, *56*, 255–264.

36. Cancio, I.; Völkl, A.; Beier, K.; Fahimi, H.D.; Cajaraville, M.P. Peroxisomes in molluscs, characterization by subcellular fractionation combined with western blotting, immunohistochemistry, and immunocytochemistry. *Histochem. Cell Biol.* **2000**, *113*, 51–60. [CrossRef]

37. Owen, G. Lysosomes, peroxisomes and bivalves. *Sci. Prog.* **1972**, *60*, 299–318.

38. Lobo-da-Cunha, A.; Batista, C.; Oliveira, E. The peroxisomes of the hepatopancreas in marine gastropods. *Biol. Cell* **1994**, *82*, 67–74. [CrossRef]

39. Lobo-da-Cunha, A.; Oliveira, E.; Alves, A.; Calado, G. Giant peroxisomes revealed by a comparative microscopy study of digestive gland cells of cephalaspidean sea slugs (Gastropoda, Euopisthobranchia). *J. Mar. Biol. Assoc.* **2019**, *99*, 197–202. [CrossRef]

40. Triebskorn, R. Ultrastructural changes in the digestive tract of *Deroceras reticulatum* (Müller) induced by a carbamate molluscicide and by metaldehyde. *Malacologia* **1989**, *31*, 141–156.

41. Lobo-da-Cunha, A.; Batista-Pinto, C. Stomach cells of *Aplysia depilans* (Mollusca, Opisthobranchia): A histochemical, ultrastructural, and cytochemical study. *J. Morphol.* **2003**, *256*, 360–370. [CrossRef] [PubMed]

42. Lobo-da-Cunha, A.; Batista-Pinto, C. Light and electron microscopy studies of the oesophagus and crop epithelium in *Aplysia depilans* (Mollusca, Opisthobranchia). *Tissue Cell* **2005**, *37*, 447–456. [CrossRef] [PubMed]

43. Lobo-da-Cunha, A.; Batista-Pinto, C. Ultrastructural, histochemical and cytochemical characterisation of intestinal epithelial cells in *Aplysia depilans* (Gastropoda, Opisthobranchia). *Acta Zool.* **2007**, *88*, 211–221. [CrossRef]

44. Tandler, B.; Phillips, C.J. Structure of serous cells in salivary glands. *Microsc. Res. Tech.* **1993**, *26*, 32–48. [CrossRef]
45. Boer, H.H.; Kits, K.S. Histochemical and ultrastructural study of the alimentary tract of the freshwater snail *Lymnaea stagnalis*. *J. Morphol.* **1990**, *205*, 97–111. [CrossRef]
46. Franchini, A.; Ottaviani, E. Intestinal cell types in the freshwater snail *Planorbarius corneus*: Histochemical, immunocytochemical and ultrastructural observations. *Tissue Cell.* **1992**, *24*, 387–396. [CrossRef]
47. Pfeiffer, C.J. Intestinal ultrastructure of *Nerita picea* (Mollusca: Gastropoda), an Intertidal marine snail of Hawaii. *Acta Zool.* **1992**, *73*, 39–47. [CrossRef]
48. Lobo-da-Cunha, A.; Malheiro, A.R.; Alves, A.; Oliveira, E.; Coelho, R.; Calado, G. Histological and ultrastructural characterisation of the stomach and intestine of the opisthobranch *Bulla striata* (Heterobranchia: Cephalaspidea). *Thalassas* **2011**, *27*, 61–75.
49. Tandler, B. Structure of mucous cells in salivary glands. *Microsc. Res. Tech.* **1993**, *26*, 49–56. [CrossRef] [PubMed]
50. Lobo-da-Cunha, A.; Oliveira, E.; Ferreira, I.; Coelho, R.; Calado, G. Histochemical and ultrastructural characterization of the posterior esophagus of *Bulla striata* (Mollusca, Opisthobranchia). *Microsc. Microanal.* **2010**, *16*, 688–698. [CrossRef] [PubMed]
51. Leal-Zanchet, A.M. Ultrastructure of the supporting cells and secretory cells of the alimentary canal of the slugs, *Lehmannia marginata* and *Boettgerilla pallens* (Pulmonata: Stylommatophora: Limacoidea). *Malacologia* **2002**, *44*, 223–239.
52. Shan, M.M.; Gentile, M.; Yeiser, J.R.; Walland, A.C.; Bornstein, V.U.; Chen, K.; He, B.; Cassis, L.; Bigas, A.; Cols, M.; et al. Mucus enhances gut homeostasis and oral tolerance by delivering immunoregulatory signals. *Science* **2013**, *342*, 447–453. [CrossRef]
53. Portela-Gomes, G.M.; Grimelius, L.; Bergström, R. Enterochromaffin (argentaffin) cells of the rat gastrointestinal tract. An ultrastructural study. *Acta Pathol. Microbiol. Immunol. Scand. A* **1984**, *92*, 83–89. [CrossRef]
54. Wade, P.R.; Westfall, J.A. Ultrastructure of enterochromaffin cells and associated neural and vascular elements in the mouse duodenum. *Cell Tissue Res.* **1985**, *241*, 557–563. [CrossRef]
55. Latorre, R.; Sternini, C.; De Giorgio, R.; Greenwood-Van Meerveld, B. Enteroendocrine cells: A review of their role in brain-gut communication. *Neurogastroenterol. Motil.* **2016**, *28*, 620–630. [CrossRef]
56. Guo, X.; Lv, J.; Xi, R. The specification and function of enteroendocrine cells in *Drosophila* and mammals: A comparative review. *FEBS J.* **2021**, *epublication ahead of print*. [CrossRef]
57. Plisetskaya, E.; Kazakov, V.K.; Soltitskaya, L.; Leibson, L.G. Insulin-producing cells in the gut of freshwater bivalve mollusks *Anodonta cygnea* and *Unio pictorum* and the role of insulin in the regulation of their carbohydrate metabolism. *Gen. Comp. Endocrinol.* **1978**, *35*, 133–145. [CrossRef]
58. Marchand, C.R.; Colard, C. Presence of cells and fibers immunoreactive toward antibodies to different peptides or amine in the digestive tract of the snail *Helix aspersa*. *J. Morphol.* **1991**, *207*, 185–190. [CrossRef]
59. Franchini, A.; Rebecchi, B.; Fantin, A.M. Immunocytochemical detection of endocrine cells in the gut of *Viviparus ater* (Mollusca, Gastropoda). *Eur. J. Histochem.* **1994**, *38*, 237–244. [CrossRef] [PubMed]
60. Cherif-Feildel, M.; Heude Berthelin, C.; Adeline, B.; Rivière, G.; Favrel, P.; Kellner, K. Molecular evolution and functional characterisation of insulin related peptides in molluscs: Contributions of *Crassostrea gigas* genomic and transcriptomic-wide screening. *Gen. Comp. Endocrinol.* **2019**, *271*, 15–29. [CrossRef] [PubMed]
61. Alba, Y.; Villaro, A.C.; Sesma, P.; Vázquez, J.J.; Abaurrea, A. Gut endocrine cells in the snail *Helix aspersa*. *Gen. Comp. Endocrinol.* **1988**, *70*, 363–373. [CrossRef]
62. Bohórquez, D.V.; Samsa, L.A.; Roholt, A.; Medicetty, S.; Chandra, R.; Liddle, R.A. An enteroendocrine cell-enteric glia connection revealed by 3D electron microscopy. *PLoS ONE* **2014**, *9*, e89881. [CrossRef] [PubMed]
63. Bohórquez, D.V.; Shahid, R.A.; Erdmann, A.; Kreger, A.M.; Wang, Y.; Calakos, N.; Wang, F.; Liddle, R.A. Neuroepithelial circuit formed by innervation of sensory enteroendocrine cells. *J. Clin. Investig.* **2015**, *125*, 782–786. [CrossRef] [PubMed]

Journal of
*Marine Science
and Engineering*

Article

How Do Prostomial Sensory Organs Affect Brain Anatomy? Phylogenetic Implications in Eunicida (Annelida)

Sabrina Kuhl *, Thomas Bartolomaeus and Patrick Beckers

Institute of Evolutionary Biology and Ecology, University of Bonn, An der Immenburg 1, 53121 Bonn, Germany
* Correspondence: skuhl@evolution.uni-bonn.de

Abstract: Eunicida is a taxon of marine annelids currently comprising the taxa Eunicidae, Onuphidae, Dorvilleidae, Oenonidae, Lumbrineridae, Histriobdellidae and Hartmaniella. Most representatives are highly mobile hunters sharing the presence of a sophisticated nervous system but differ in the number and shape of prostomial sensory organs (0–3 antennae; 0 or 2 palps; 0, 2 or 4 (+2) buccal lips; 0, 2 or 4 eyes; single-grooved or paired nuchal organs). This makes Eunicida an ideal model to study the following questions: Is the brain morphology affected by different specificities of prostomial sensory organs? Do similar numbers and shapes of prostomial sensory organs hint at close phylogenetic relationships among different eunicidan taxa? How can antennae, palps and buccal lips be differentiated? For the investigation of sensory organs and the nervous system, we performed immunohistochemistry, μCT, TEM, SEM, paraffin histology and semi-thin sectioning. Our results show that brain anatomy is mostly affected on a microanatomical level by sensory organs and that similar specificities of sensory organs support the latest phylogenetic relationships of Eunicida. Further, a reduction of antennae in Eunicida can be suggested and hypotheses about the presence of sensory organs in the stem species of Eunicida are made.

Keywords: neuropil; nerve; reduction; nuchal organ; antenna; palp; eye; histology; electron microscopy

Citation: Kuhl, S.; Bartolomaeus, T.; Beckers, P. How Do Prostomial Sensory Organs Affect Brain Anatomy? Phylogenetic Implications in Eunicida (Annelida). *J. Mar. Sci. Eng.* **2022**, *10*, 1707. https://doi.org/10.3390/jmse10111707

Academic Editor: Alexandre Lobo-da-Cunha

Received: 31 August 2022
Accepted: 7 November 2022
Published: 9 November 2022

Publisher's Note: MDPI stays neutral with regard to jurisdictional claims in published maps and institutional affiliations.

1. Introduction

Eunicida currently comprises over 1000 species in the major taxa Eunicidae (Berthold, 1827), Onuphidae (Kinberg, 1865), Dorvilleidae (Chamberlin, 1919), Oenonidae (Kinberg, 1865), and Lumbrineridae (Schmarda, 1861) and the two minor subtaxa Histriobdellidae (Claus and Moquin-Tandon, 1884) and Hartmaniella (Imajima, 1977) [1–6]. Fossils date Eunicida back to the late Cambrian [6–9], underlining the great diversification this clade encountered over time [7]. All members of Eunicida are exclusively marine, conquering habitats ranging from shallow coastal lines to deep sea areas [7]. Most representatives of Eunicida have a predatory lifestyle: they hunt their prey actively down with their complex jaw apparatus [6,8,10]. Interaction with the environment and perception of chemical cues are performed by a set of diverse sensory organs such as head appendages (antennae and palps), buccal lips, pigment cup-shaped cerebral eyes, nuchal organs and lateral organs. The number of head appendages and eyes shows extreme variation between the different families of Eunicida and becomes visible even on the family level [3,5,11,12]: members of the family Eunicidae for example comprise species bearing five, three or one head appendage(s) and one pair of eyes, while members of the family Lumbrineridae lack head appendages as well as eyes completely. In between these extremes, diverse species-specific combinations of prostomial sensory organs are present.

All eunicidan prostomial sensory organs are innervated by nerves originating in the brain neuropil [11,13,14]. The brain is located in the prostomium and constitutes the central nervous system together with the circumoesophageal connectives and the ventral nervous system [10,13,15]. In the brain, neurites are frequently arranged in bundles with certain orientations, termed tracts or commissures [11,13,14,16]. Tracts and commissures serve as connectors, e.g., between nerves innervating sensory organs, so that different neuronal innervation patterns may become apparent among brains of different eunicidan taxa.

Since the 19th century, scientists have tried to describe the variety of head appendages in Eunicida using a set of different terms. Over time, the terminology changed according to new insights (homology hypotheses) and the application of new techniques (cLSM). A selection of these terms is given in Table S2. If compared to the latest nomenclature used, there is a historical consensus in terming the median and lateral antennae as "antennae". Solely the studies of Monro [17] and Aiyar [18] are an exception to this in using the term "tentacle".

Historical nomenclature of structures nowadays accepted as palps passed through greater changes. Many authors—former and present ones—use the term "antennae" to describe palps. Presumably led by the stunning same outer morphology of antennae and palps in most eunicidan taxa on the one-hand side (e.g., Figure 1) and prominent differences of eunicidan palps with, e.g., feeding and sensory palps of other annelid taxa on the other. With the advance of confocal laser scanning microscopy and generally improved imaging methods, the idea of differentiation by innervation patterns was put forward. Since palps in all annelid taxa are innervated by both the drcc and the vrcc and their commissures, this criterium served as defining feature for palps in Eunicida for the first time in the study of Orrhage and Müller [13]. Interestingly, former authors such as Hanström [12], Fauchald [19] and Fauchald and Rouse [20] used the term "palp" prior to this definition for head appendages in the "ventro-lateral" position in Dorvilleidae.

Buccal lips were described generally as the "anterior-most outgrowth" of the prostomium and similar to the classification of palps, many different terms were used to describe buccal lips over time. Most of the authors regarding all five head appendages as antennae termed buccal lips "palps", e.g., Ehlers [5], Heider [21], Hanström [12], von Haffner [22,23], Paxton [24] and Orensanz [25]. The ventro-lateral antennae of Eunicidae and Onuphidae were homologized with the palps of Nereididae and Aphroditidae by Binard and Jeener [26] and Gustafson [27], who termed the eunicidan buccal lips consequently "faux-palpes". Aiyar [18] who neither used the terms "antenna" nor "palp", named the buccal lips neutrally "lobes". This terminology for buccal lips was adopted by Åkesson [28] in labelling them "oral lobes". Since Orrhage [14] introduced the term "buccal lip", a consensus in the labelling of buccal lips exists. No homology of eunicidan buccal lips and palps of other annelid taxa was suggested by Binard and Jeener [24] and Orrhage [14]. The same conclusions were drawn by Zanol [29] due to differing innervation patterns. The dorsal buccal lips found in some taxa of Onuphidae passed through an intensive phase of renaming, too. From "palps" [12,22–24,30] over "antennae" [5,19,31] to the now-accepted term "ventral buccal lip" [6,13,14,20,29,32].

The differing numbers of head appendages make Eunicida an ideal model to study the following questions:

1. Is the brain morphology affected by different numbers and shapes of prostomial sensory organs? It could be assumed, that species with many prostomial sensory organs possess complex innervation patterns for processing of stimuli, while species lacking eyes and head appendages may show a more simple brain morphology. However, it could also be hypothesized, that species with a less amount of prostomial sensory structures balance missing sensory input out by expanding another sense (e.g., chemosensation of nuchal organs) as von Haffner [11] already suggested.
2. Do similar numbers and shapes of prostomial sensory organs hint at close phylogenetic relationships among different eunicidan taxa?
3. How can head appendages (antennae and palps) as well as buccal lips be differentiated?

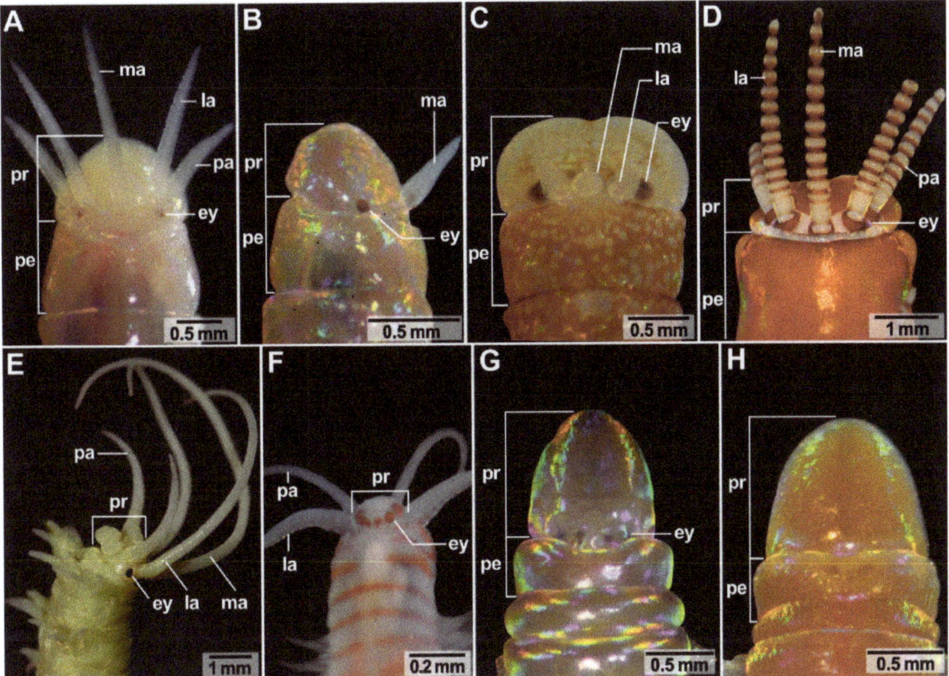

Figure 1. Variation of head appendages (antennae, palps) and eyes in representatives of Eunicida. Living specimens (**A–D,F–H**), fixated specimens (**E**). (**A**) *Paucibranchia bellii*, dorsal view. Five head appendages—one median antenna (ma), one pair of lateral antennae (la) and one pair of palps (pa) are present. One pair of small red-pigmented eyes (ey) is located at the lateral basis of the lateral antennae. (**B**) *Lysidice unicornis*, lateral view. One median antenna (ma) and one pair of red-pigmented eyes (ey) are located on the prostomium. (**C**) *Lysidice ninetta*, dorsal view. Three head appendages are present (1 ma,1 pair of la) and one pair of relatively big red-pigmented eyes is located on the ventral basis of the lateral antennae. (**D**) *Leodice torquata*, dorsal view. Five head appendages are present (1 ma, 1 pair of la, 1 pair of pa). One pair of blackish-purple eyes (ey) is located on the ventral basis of the lateral antennae. (**E**) *Hyalinoecia tubicola*, lateral view. Five relatively long head appendages are present (1 ma, 1 pair of la, 1 pair of pa). One pair of black-pigmented eyes (ey) is located at the basis of the lateral antennae. The specimen was already fixed when the picture was taken—deviation in colouration of the specimen is possible. (**F**) *Dorvillea* spec., dorsal view. Four relatively long head appendages are present (1 pair of la, 1 pair of pa). Two pairs of red-pigmented eyes (ey) are located on the posterior edge of the prostomium. (**G**) *Arabella iricolor*, dorsal view. Head appendages are lacking, but two pairs of black-pigmented eyes (ey) are present on the prostomium. (**H**) *Scoletoma tetraura*, dorsal view. Lacks head appendages as well as eyes. ey = eye, la = lateral antenna, ma = median antenna, pa = palp, pe = peristomium, pr = prostomium.

To provide answers to these research questions, we analyzed the innervation of sensory organs and reconstructed their tracts and commissures in the brain in a broad comparative morphological approach (paraffin histology, immunohistochemistry, ultrastructural analyses, scanning electron microscopy, micro-computed tomography and 3D reconstructions). Further, we compared the morphology of antennae, palps, buccal lips, eyes, and nuchal organs among the different taxa to evaluate if interspecific differences are present. These differences might explain possible variations in the innervation of sensory organs on the one hand, while on the other hand, they may serve as indicators for hypotheses of phylogenetic relationships in Annelida. Generally, morphological studies are still crucial to understand the evolution of divergent morphological traits, especially in taxa as diverse as Annelida. A mapping of morphological characters on recent molecular phylogenies of Eunicida provides a reliable base for the formulation of phylogenetic and evolutionary hypotheses, shedding new light on old evolutionary questions.

2. Materials and Methods

2.1. Animals

Specimens of *Leodice* (*Eunice*) *torquata* (Quatrefages, 1866), *Lysidice* (*Nematonereis*) *unicornis* (Grube, 1840) were collected in the intertidal zone of Le Cabellou in Concarneau (France, Brittany) in March 2017 (Table S1, Figure 1). Specimens of *Scoletoma* (*Lumbrineris*) *tetraura* (Schmarda, 1861), *Lysidice ninetta* (Audouin and Milne Edwards, 1833), *Paucibranchia* (*Marphysa*) *bellii* (Audouin and Milne Edwards, 1833) and *Marphysa* spec. were found in the intertidal zone in Concarneau (France, Brittany) in April 2017 (Table S1, Figure 1). *Hyalinoecia tubicola* (O.F. Müller, 1776) was collected in Bergen (Norway) in August 2017 (Table S1). *Diopatra neapolitana* (Delle Chiaje, 1841) was found in Zostera beds during low tide in the bay of Arcachon close to Le Petit Piquey (France, Nouvelle-Aquitaine) in 1995 (Table S1). Specimens of *Dorvillea* spec. were found in an aquarium in Leipzig (Germany) (Table S1, Figure 1). Specimens of *Ophryotrocha siberti* were collected in an aquarium in San Diego (USA) in 2015. Specimens of *Arabella iricolor* (Montagu, 1804) were collected in the intertidal zone in Le Cabellou in Concarneau (France, Brittany) in 2014 and 2015 (Table S1, Figure 1). *Scoletoma* (*Lumbrineris*) *fragilis* (O.F. Müller, 1776) was collected in Le Cabellou in Concarneau (France, Brittany). *Histriobdella homari* (Beneden, 1858) was collected from commercially bought Lobster in Bergen (Norway) in 2018 (Table S1).

2.2. Paraffin Histology

Specimens were relaxed in 7% $MgCl_2$ mixed with seawater (1:1) followed by prefixation in 7% formaldehyde in seawater (1:1). Fixation was performed in the refrigerator overnight using Bouin's fixative modified after Dubosque-Basil. The further steps applied follow the standardised methodology described by [33–35]. Serial sections (thickness 5 µm) of the animal's prostomium and the peristomium were created using a Reichert-Jung Autocut 2050 microtome (Leica, Wetzlar). Specimens were either cut in cross, horizontal or sagittal sections and were stained in AZAN (for a detailed description see Beckers et al. [35]) or impregnated with silver modified after the protocol of Palmgren (1948, 1955, 1960) (Table S1).

2.3. Immunohistochemistry

One specimen of *Scoletoma tetraura* and two specimens of *Lysidice ninetta* were relaxed in a 7% $MgCl_2$ solution mixed with seawater (1:1). Fixation was performed in a 4% paraformaldehyde solution (Electron Microscope Sciences, Hatfield, PA, USA) in seawater at 4 °C overnight. Further steps were conducted according to standardized immunohistological protocols. Samples were cut using a vibratome (Micron HM 650 V, Thermo Scientific, Dreieich, Germany) producing 60 µm thick sections. Sections were stained using FMRF-amide (rabbit; ImmunoStar, Hudson, WI, USA) at a dilution of 1:2000 and α-tubulin (mouse; Sigma-Aldrich, Germany) at a dilution of 1:500 as primary antibodies. The secondary antibodies Cy2 (goat anti-rabbit; ImmunoStar, Hudson, WI, USA) and Cy5 (goat anti-mouse; Sigma-Aldrich, Germany) were added in a dilution of 1:500 each. Sections were mounted on chromalaun/gelatine-covered glass slides and embedded with Elvanol (after Rodriguez and Deinhard (1960) modified after M. Bastmeier). Examination of the sections was performed using a Leica confocal laser scanning microscope (TCS SPE, Germany, excitation wavelength: 488 nm/635 nm) and images of the samples were taken via the software LAS AF 1.6.1 (Leica, Microsystems).

2.4. Micro-Computed Tomography (µCT)

Specimens of *Lysidice ninetta*, *Leodice torquata* and *Hyalinoecia tubicola* were investigated via micro-computed tomography (µCT) (Table S1). First, specimens were relaxed in a solution of 7% $MgCl_2$ mixed with seawater (1:1). Prefixation was performed in a 7% formaldehyde solution mixed with seawater (1:1) and fixation was performed in Bouin's fixative overnight. Samples were washed with 70% ethanol and stained in a 0.3% phosphotungstic acid (PTA) solution in 70% ethanol for two weeks. Imaging was performed using a SkyScan 1272 µCT scanner (Burker, Kontich, Belgium), which was equipped with a Hamamatsu L11871 20 tungsten X-ray source (Hamamatsu, Japan) and a Ximea xiRAY 16 camera (XIMEA GmbH, Germany). The following scanning parameters were used: source voltage = 70 kV, source current = 142 µA, exposure time = 1430 ms, frames averaged = 10, flat field correction = activated and scanning time = 14–20 h.

2.5. Electron Microscopy

For scanning electron microscopy (SEM) trunk segments of *Paucibranchia bellii*, *Marphysa* spec., *Lysidice unicornis*, *Lysidice ninetta*, *Leodice torquata*, *Hyalinoecia tubicola*, *Diopatra neapolitana*, *Ophryotrocha siberti*, *Arabella iricolor* and *Scoletoma tetraura* were investigated focusing on the lateral organs (Table S1). Fixation and further processing of the samples were conducted according to the methodology described by Beckers et al. [33]. Dehydration was performed in an ascending acetone series if followed by critical-point drying (Bal-Tec CPD 030, Switzerland) or via an ascending ethanol series if treated with HMDS (hexamethyldisilazane). After mounting on aluminium stubs, samples were sputter coated with gold (SEM coating unit E5100, Polaron Equipment Ltd., Great Britain) and were examined via a scanning electron microscope (Leitz AMR 1000, FEI Verios 460 L).

The anterior part of one specimen of *Paucibranchia bellii* and *Scoletoma fragilis* was investigated via semi-thin sections using light microscopy (BX-51, Olympus). Additionally, *Paucibranchia bellii* was examined via ultra-thin sections in transmission electron microscopy (TEM). Fixation and further processing of the samples were performed according to the methodology described by Beckers et al. [33]. Then, 1 µm semi-thin and 70 nm thick ultra-thin sections were created using a Leica EM Ultracut 6 microtome and a diamond knife. Semi-thin sections were transferred to glass slides and stained with Toluidine blue. Ultra-thin sections were placed on formvar-coated single slot copper grids (1 × 2 mm) and stained with uranyl acetate and lead citrate automatically (QG-3100, Boeckler Instruments). Examination of the ultra-thin sections was conducted using a transmission electron microscope (Zeiss EM 10CR). Negative pictures documented on phosphor imaging plates were digitalized using an imaging plate scanner (Ditabis).

2.6. Data Analyses and 3D Reconstructions

Living specimens were photographed using a Canon 600D camera mounted on a Zeiss-Stemi 2000. Paraffin and semi-thin sections were analysed with an Olympus microscope (BX-51) that was equipped with an Olympus camera (Olympus cc12) and the dot slide system (2.2 Olympus, Hamburg). Automatic alignment of the images was performed using imod and imodalign (http://www.q-terra.de/biowelt/3drekon/guides/imod_first_aid.pdf, accessed on 5 October 2022). For the creation of an image stack the images were loaded in ImageJ 1.51 k (Wayne Rasband, National Institutes of Health, USA, http://imagej.nih.gov/ij, accessed on 5 October 2022) and segmentation was performed using TrackEM in ImageJ. The 3D reconstructions were created in Amira 5.3.1 (Thermo Fisher Scientific, Waltham, MA, USA).

Z-projections of the immunohistological image stacks were created using ImageJ 1.51 k and the function "maximum intensity". If images needed to be stitched together, 3D stitching was performed in ImageJ 1.51 k.

Two-dimensional (2D) image slicing of the µCT data was performed using ImageJ 1.51 k and the plug-in Volume Viewer. The 3D rendering was conducted in Amira 5.3.1

and 3D reconstructions of the nervous systems of the samples were performed by manual segmentation and surface rendering.

Image processing and creation of schematic drawings were conducted in Adobe Photoshop (CS6) and Adobe Illustrator (CS6).

3. Results

The neuroanatomical nomenclature used in this study follows the terminology given by Richter et al. [36]. Without proposing a general homology with palps of other annelids, the term "palp" is used here to describe head appendages innervated by two or more nerve roots originating from the dorsal root of the circumoesophageal connective (drcc) and the ventral root of the circumoesophageal connective (vrcc) and their commissures [13]. According to Orrhage and Müller [13] and Orrhage [14], antennae are solely innervated by nerves emanating from the drcc. Based on the study of Zanol [29], we use the adjective "eunicid" in context with all members belonging to the family Eunicidae, while "eunicidan" refers to members of the order Eunicida.

3.1. The Central Nervous System

The central nervous system (cns) of Eunicida is basiepidermal (intraepidermal) and consists of a central neuropil (brain), the circumoesophageal connectives (cc) and the paired ventral nerve chord (vnc) (Figure 2A–F). The brain is located dorsally in the prostomium and is connected via the cc in the peristomium to the suboesophageal ganglion and the subsequent paired ventral nerve chord in the metameric trunk (Figure 2A–F). The cc splits in a dorsal (drcc) and a ventral root (vrcc) in the brain and the ganglionic ventral nervous system forms regular segmental commissures running through the trunk (Figure 2A–F).

Comparing brain positions among the different species studied, the brain of *Arabella iricolor* (Oenonidae), *Scoletoma tetraura* and *Scoletoma fragilis* (Lumbrineridae) is positioned a bit more posteriorly in the prostomium than in the remaining taxa (Figure 2G,H). This impression might be consolidated by the elongated prostomium of these taxa.

3.2. General Brain Anatomy

In all species investigated, the brain surrounds the intracerebral cavity (Figure 3A,C,F,G). Anteriorly to this point, the brain splits into smaller neuropils, from which many delicate nerves arise innervating the tip of the prostomium (Figure 2G,H). At the height of the intracerebral cavity, dorsal and ventral brain neuropil can be differentiated with regard to their relative topology (Figure 3A,C,F,G). The dorsal brain neuropil innervates all dorsally located sensory organs (such as antennae, eyes and nuchal organs) (Figure 3C,D,H,I) and inhabits the respective commissures and tracts, e.g., optical commissures and association tracts of antennae. Parts of the ventral brain neuropil innervate the buccal lips—if present. Palps are innervated by both, the dorsal and the ventral brain neuropil. The oesophageal ganglion is still part of the ventral brain neuropil and gives rise to the jaw nerves (Figure 3D,E,H,I). The dorsal root of the circumoesophageal connective (drcc) originates from the dorsal brain neuropil (Figure 3D,I). The ventral brain neuropil gives rise to the ventral root of the circumoesophageal connective (vrcc) (Figure 3I). In many species, e.g., in *Lysidice ninetta*, the dorsal and the ventral commissure of the ventral root of the circumoesophageal connective (dcvr/vcvr) can be differentiated (Figure 2D). With this, the ventral brain neuropil terminates. The dorsal brain neuropil proceeds posteriorly and innervates the nuchal organ before it terminates (Figure 2G).

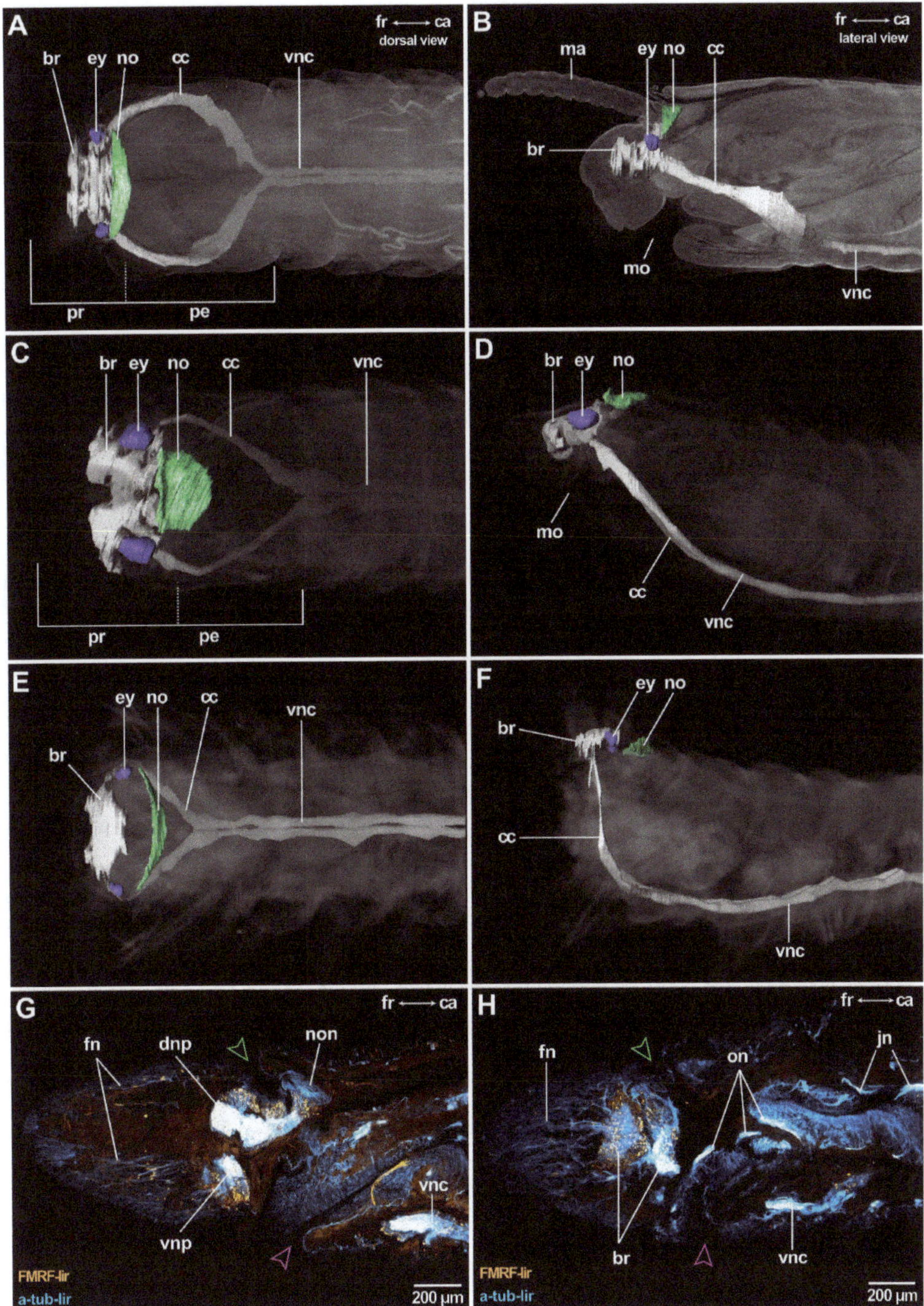

Figure 2. Components and topology of the cns in Eunicida. The 3D reconstructions and volume renderings are based on µCT scans (**A–F**), cLSM scans of sagittal sections (**G,H**). (**A**) *Leodice torquata*, dorsal view. (**B**) *L. torquata*, lateral view. (**C**) *Lysidice ninetta*, dorsal view. (**D**) *L. ninetta*, lateral view.

(**E**) *Hyalinoecia tubicola*, dorsal view. (**F**) *H. tubicola*, lateral view. (**G**) *Scoletoma tetraura*, lateral view. A fine net of frontal nerves (fn) originating from the brain neuropil innervates the anterior part of the prostomium. The dorsal and the ventral brain neuropil (dnp, vnp) can be distinguished. Nuchal organ nerves (non) originate from the posterior-most part of the dorsal brain neuropil (dnp) and innervate the nuchal organ (indicated by green arrowhead). (**H**) *S. tetraura*, lateral view. From the ventral brain neuropil, the oesophageal nerves (on) originate at height of the mouth opening (purple arrowhead). br: brain; ca: caudal; cc: circumoesophageal connective; cns: central nervous system; dnp: dorsal brain neuropil; ey: eye; fn: frontal nerves; fr: frontal; jn: jaw nerves; ma: median antenna; mo: mouth opening; no: nuchal organ; non: nuchal organ nerve; on: oesophageal nerves; pe: peristomium; pr: prostomium; vnc: ventral nerve chord; vnp: ventral brain neuropil; green arrow: nuchal cavity; pink arrow: mouth opening.

Deviations from this brain morphology occur in *Arabella iricolor* and *Scoletoma tetraura*. In both species, the anteriorly located neuropil parts extend much more ventrally than in all other species examined (Figure 3H). In *A. iricolor*, the oesophageal ganglion originates from the ventral parts of these neuropils (Figure 3H,I). The oesophageal ganglion is connected to the circumoesophageal connective via a pair of distinct commissures (Figure 3G,I). Both the commissures as well as the oesophageal ganglion are surrounded by a thick layer of intermediate filaments (Figure 3G). The paired stomatogastric nerves originate from these commissures innervating the stomatogastric system including the oesophagus (Figure 3J). This situation was not observed in any other species examined. Further, the intracerebral cavity in *A. iricolor* is much smaller than in any other taxon investigated (Figure 3F,H). In *S. tetraura* the ventrally extended anterior neuropil parts fuse in the posterior course so that together with the dorsal brain neuropil it surrounds the intracerebral cavity.

Additionally, in *A. iricolor*, as well as *S. tetraura*, distinct neuropils occur at the lateral sides of the brain neuropil. In both species, these lateral neuropils extend dorso-ventrally and they are surrounded by a thick layer of somata of type 1 neurons (Figure 4A,D). The lateral neuropils are interconnected by a tract located in the dorsal brain neuropil (Figures 3F,G and 4B,C,E). The tract is surrounded by intermediate filaments of glia cells (Figures 3B,F and 4C). In *A. iricolor*, the dorsal nerves root in this tract and the dorsal brain neuropil (Figure 4C).

3.3. Ultrastructure of the Central Neuropil and the Somata

In all species investigated, the central neuropil consists of neurites and is surrounded by somata of different neuronal types, that are discriminated by size (Figure 5A–C,E). The neurites can have different diameters and orientations and parallel-oriented neurite bundles create commissures and tracts in the otherwise heterogeneous central neuropil (Figure 5A,D). Dense and lucent core vesicles comprising neuropeptides are present in the neurites (Figure 5F) and stabilizing intermediate filaments of glial cells surround the neurites (Figure 5D). Blood vessels may traverse the brain. A thin layer of glial cells and their processes isolate the central neuropil from the surrounding somata (Figure 5A,D). Each soma is isolated by several layers of glia cell processes (Figure 5D) while desmosomes connect adjacent glia cell processes (Figure 5D). A glial layer surrounding the neuronal somata is not visible (Figure 5A).

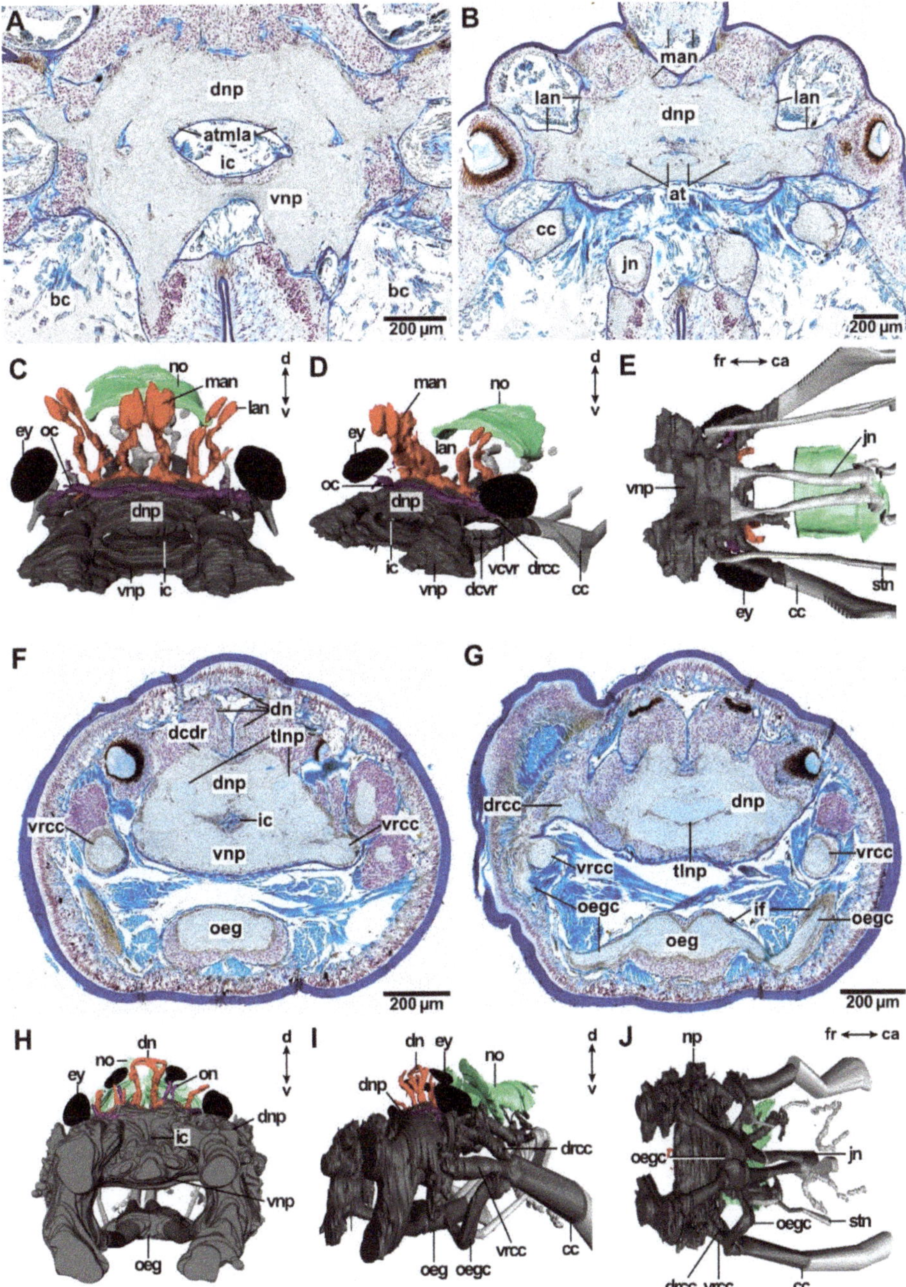

Figure 3. General brain anatomy. Histological cross sections (5 μm, AZAN) of *Leodice torquata* (**A**,**B**) and *Arabella iricolor* (**F**,**G**). Three-dimensional reconstructions based on histological serial sections of *Lysidice ninetta* (**C**–**E**) and *A. iricolor* (**H**–**J**). at: association tract; atmla: association tracts of the median and the lateral antennae; bc: buccal cavity; ca: caudal; cc: circumoesophageal connective; d: dorsal; dcvr: dorsal commissure of the ventral root of the circumoesophageal connective; dn: dorsal nerves;

dnp: dorsal brain neuropil; drcc: dorsal root of the circumoesophageal connective; ey: eye; fr: frontal; ic: intracerebral commissure; if: intermediate filaments; jn: jaw nerves; lan: lateral antenna nerve; man: median antenna nerve; no: nuchal organ; oc: optical commissure; oegc: connective of the oesophageal ganglion; oen: oesophageal nerves; on: optical nerve; stn: stomatogastric nerves v: ventral; vcvr: ventral commissure of the ventral root of the circumoesophageal connective; vnp: ventral brain neuropil; vrcc: ventral root of the circumoesophageal connective.

Somata of type 1 neurons (S1,) are densely packed and their nuclei appear as reddish–purple spherical structures with a size of 4.4–7.5 μm (Figure 5A,B). S1 neurons occur predominantly on the lateral sides of the central neuropil (Figure 5A). As somata of type 2 neurons (S2) possess more glia enwrapping than S1 neurons, S2 neurons appear less dense packed. Both types of neurons share a similar-sized nucleus (Figure 5A,C). S2 neurons occur constantly around the central neuropil surrounding it dorsally, laterally as well as ventrally (Figure 5A). Somata of type 3 neurons (S3) are enlarged and their nuclei are more prominent (13.2–26.3 μm) than the nuclei of S1 and S2 neurons (Figure 5A,E). The nucleolus can easily be identified in S3 neurons due to its brightly red stain while the nucleus itself appears light blue (Figure 5E). S3 neurons do not occur constantly around the central neuropil since they are refined to certain areas: they occur medio-dorsally and ventro-laterally to the central neuropil and only at the frontal part of the dorsal neuropil (Figure 5A).

3.4. Innervation Patterns of Antennae in the Brain

Antennal innervation patterns in the brain may differ among the species studied according to the number of antennae and palps present. Most complex innervation patterns can be found in Eunicidae bearing three antennae and two palps (*Paucibranchia bellii*, *Marphysa* spec., *Leodice torquata*) in possessing so-called association tracts (sensu Orrhage 1995), while association tracts seem to be missing in all other species examined (Figure 6).

3.4.1. Innervation Patterns in Eunicidae (Five Head Appendages)

In *Pauchibranchia bellii*, *Marphysa* spec. and *Leodice torquata*, antennae are innervated by one main nerve (man, lan) (Figure 7A). In the median antenna, this nerve splits into two main nerve roots (manr1) at the height of the ceratophore (Figures 7A and 8B–D). The main nerve roots run along both sides of the antennae coelom towards the brain. Each manr1 splits into three minor nerve roots (manr2, manr3, atma) prior to entering the dorsal brain neuropil (Figure 7A). The manr2 and manr3 roots in the dorsal commissure of the dorsal root of the circumoesophageal connective (dcdr). The third minor nerve root is termed the association tract (atma) according to Orrhage (1995) and runs ventrally through the dorsal brain neuropil (Figures 7A and 8A,D,E,G). Both association tracts of the median antenna (atma) fuse in the dorsal neuropil (Figures 7A and 8E,G), and split again at the ventral side of the dorsal brain neuropil (Figures 7A and 8B–D). There, each atma fuses with the respective association tract of the lateral antennae (atla) (Figures 7A and 8B,D). The product of this fusion is the paired association tracts of the median and the lateral antennae (atmla) that proceed next to the intracerebral cavity in the anterior direction through the dorsal brain neuropil (Figures 7A and 8D). In the anteriorly located part of the brain the paired atmla fuse in a larger area of intensively stained neurites. In this area, the dorsal roots of the palp nerves originate.

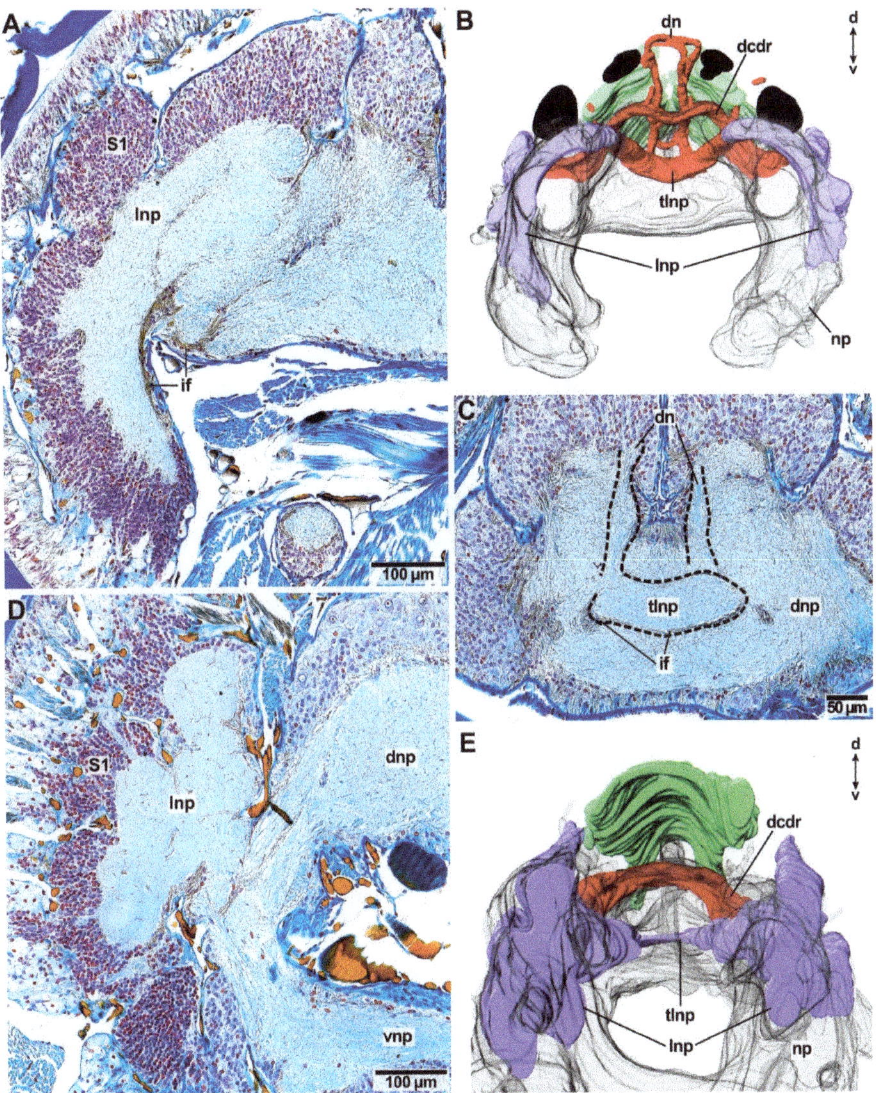

Figure 4. Lateral brain neuropils in *Arabella iricolor* (Oenonidae) and *Scoletoma tetraura* (Lumbrineridae). Histological cross sections (5 µm, AZAN) (**A,C,D**). Three-dimensional reconstructions based on histological serial sections (**B,E**). (**A**) The lateral brain neuropil (lnp) is surrounded by a dense layer of somata of type 1 neurons (S1) and intermediate filaments (if) of glia cells in *A. iricolor*. (**B**) 3D reconstruction of the lnp and the tract (tlnp) connecting both neuropils in the brain (np) in *A. iricolor*. The dorsal nerves (dn) root partially in the tlnp. Frontal view. (**C**) The dn root in the tlnp and the dorsal brain neuropil (dnp) in *A. iricolor*. The tract is surrounded by intermediate filaments (if) of glia cells. (**D**) A dense layer of somata of type 1 neurons (S1) surround the lnp in *S. tetraura*. (**E**) A 3D reconstruction of the lnp and the tlnp connecting both neuropils in the brain (np) of *S. tetraura*. Frontal view. d: dorsal; dcdr: dorsal commissure of the dorsal root of the circumoesophageal connective; dn: dorsal nerves; dnp: dorsal brain neuropil; if: intermediate filaments; lnp: lateral brain neuropil; np: neuropil; S1: somata of type 1 neurons; tlnp: tract of the lateral brain neuropils; v: ventral; vnp: ventral brain neuropil.

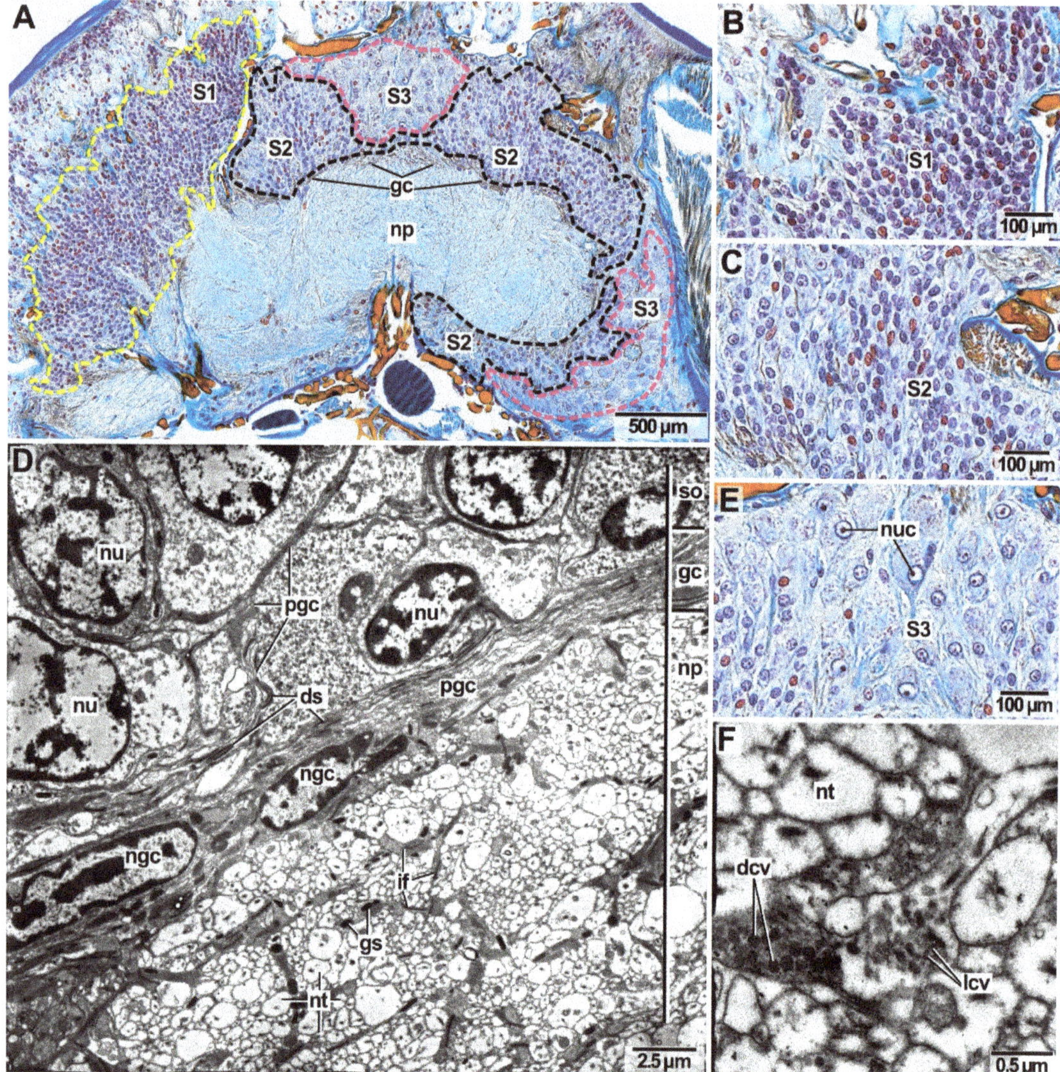

Figure 5. Brain neuropil and different types of neuronal somata. Ultra-thin cross sections (70 nm) of *Paucibranchia bellii* (**A**,**B**) and histological cross sections (5 μm, AZAN) of *Scoletoma tetraura* (**C**–**F**). (**A**) Distribution of different types of somata (S1, S2, S3) around the brain neuropil (np). (**B**) Somata of type 1 neurons (S1). (**C**) Somata of type 2 neurons (S2). (**D**) Somata of type 3 neurons (S3) with prominent nucleoli (nuc) in the nucleus. © The brain neuropil (np) consists of neurites (nt) of different sizes and orientations. Processes of glial cells (pgc) surround the central neuropil (np) creating a layer between the central neuropil (np) and the neuronal somata (so). Nuclei of the glial cells (ngc) are brick-shaped and processes of the glial cells (pgc) can be distinguished. Neuronal somata (so) are interconnected by desmosomes (ds). (**F**) Dense (dcv) and lucent core vesicles (lcv) are present in the brain containing neurotransmitters. dcv: dense core vesicles; ds: desmosomes; gc: glia cells; if: intermediate filaments; lcv: lucent core vesicles; ngc: nucleus of glia cell; np: neuropil; nt: neurite; nu: nucleus; nuc: nucleolus; pgc: processes of glia cell; S1: somata of type 1 neurons; S2: somata of type 2 neurons; S3: somata of type 3 neurons; so: somata.

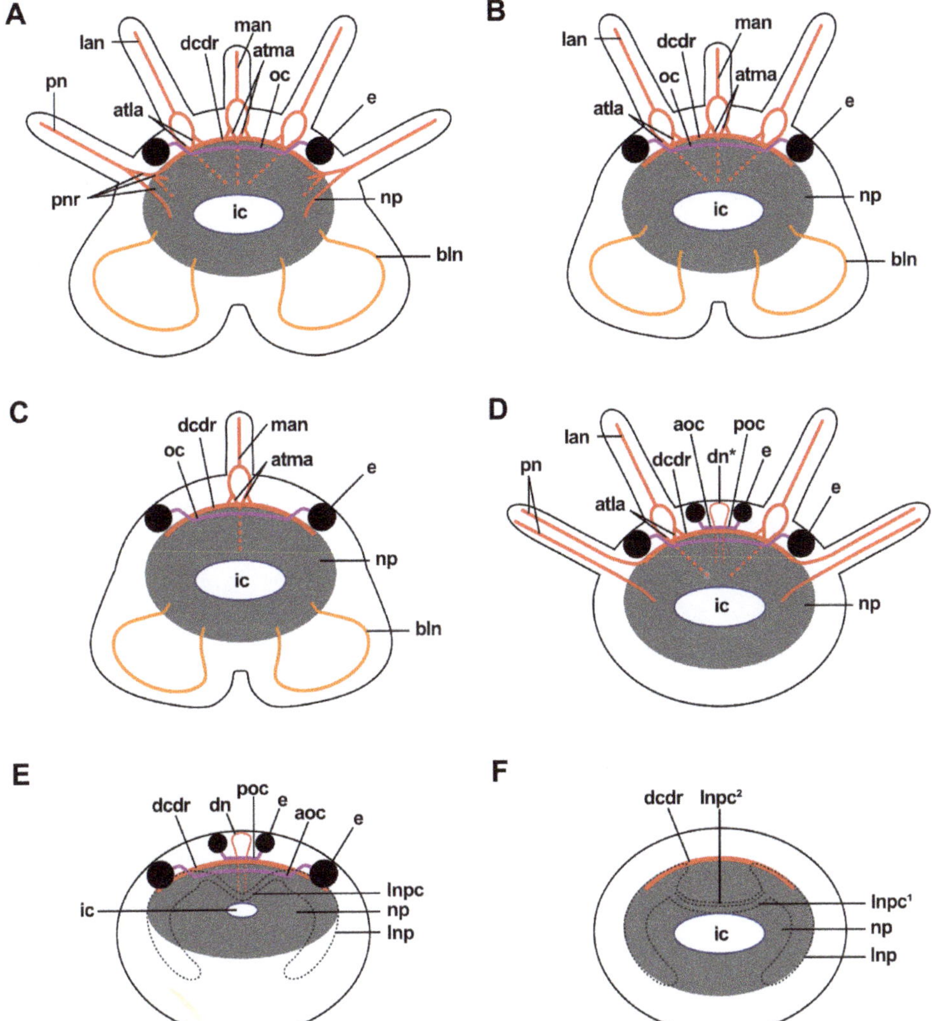

Figure 6. Generalized schemes of the innervation patterns of species bearing different numbers of antennae, palps, and eyes and conditions in species lacking these sensory organs. The asterisk (*) in picture D indicates the inclusion of data from Zanol [20]. (**A**) Innervation patterns found in Eunicidae and Onuphidae bearing 5 head appendages (*Pauchibranchia bellii*, *Marphysa* spec., *Leodice torquata*, *Hyalinoecia tubicola*). (**B**) Innervation patterns found in *Lysidine ninetta* (3 head appendages). (**C**) Innervation patterns of *Nematonereis unicornis* (1 head appendage). (**D**) Innervation patterns of *Dorvillea* spec. (4 head appendages). (**E**) Innervation patterns of *Arabella iricolor* (lacking head appendages, but possessing two pairs of eyes). (**F**) Conditions found in *Scoletoma tetraura* and *Scoletoma fragilis* (lacking head appnendages and eyes). aoc: anterior optical commissure; atla: association tract of the lateral antennae; atma: association tract of the median antenna; bln: buccal lip nerve; dcdr: dorsal commissure of the dorsal root of the circumoesophageal connective; dn: dorsal nerve; e: eye; ic: intracerebral cavity; lan: lateral antenna nerve; lnp: lateral neuropil; lnpc: lateral neuropil commissure; lnpc1: first commissure of the lateral neuropils; lnpc2: second commissure of the lateral neuropils; man: median antenna nerve; np: brain neuropil; oc: optical commissure; pn: palp nerve; pnr: palp nerve root; poc: posterior optical commissure.

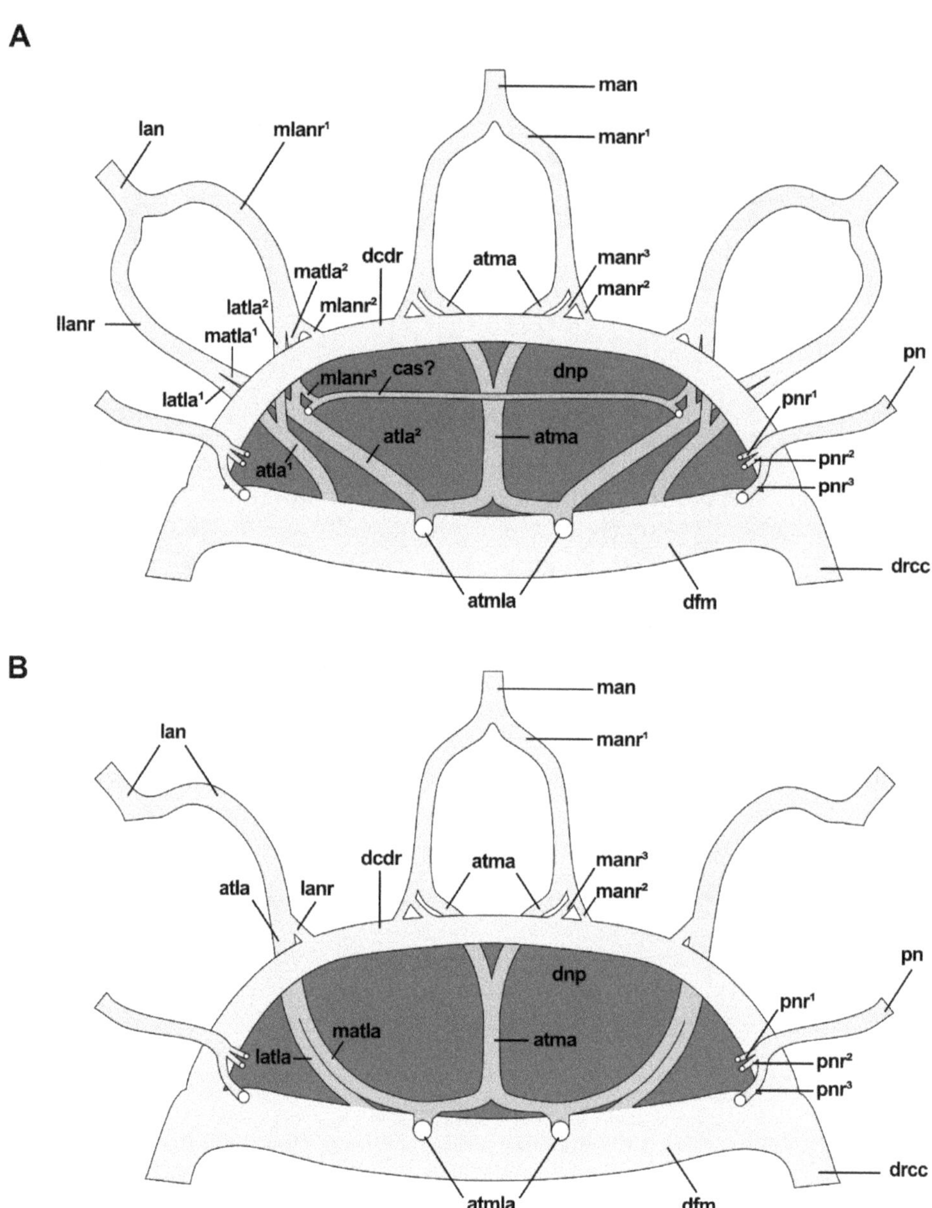

Figure 7. Schematic drawings of association tracts and nerve roots (light grey, reddish) of species bearing five head appendages based on a serial histological cross section of the dorsal brain neuropil (dark grey). (**A**) Eunicidae (*Marphysa* spec., *Paucibranchia bellii* and *Leodice torquata*). (**B**) Onuphidae (*Hyalinoecia tubicola*). atla: association tract of the lateral antenna; atla1: first association tract of the lateral antenna; atla2: second association tract of the lateral antenna; atma: association tract of the median antenna; atmla: association tract of the median and the lateral antennae; cas: commissure of the association system; dcdr: dorsal commissure of the dorsal root of the circumoesophageal connective; dfm: dorsal fibril mass; dnp: dorsal brain neuropil; drcc: dorsal root of the circumoesophageal connective; lan: lateral antenna nerve; lanr: nerve root of the lateral

antenna; latla: lateral association tract of the lateral antenna; latla1: lateral root of the first association tract of the lateral antenna; latla2: lateral root of the second association tract of the lateral antenna; llanr: lateral root of the lateral antenna nerve; man: median antenna nerve; matla: median association tract of the lateral antenna; matla1: median root of the first association tract of the lateral antenna; matla2: median root of the second association tract of the lateral antenna; manr1: first root of the median antenna nerve; manr2: second root of the median antenna nerve; manr3: third root of the median antenna nerve; mlanr: median root of the lateral antenna nerve; mlanr2: second median root of the lateral antenna nerve; mlanr3: third median root of the lateral antenna nerve; pn: palp nerve; pnr1: first root of the palp nerve; pnr2: second root of the palp nerve; pnr3: third root of the palp nerve.

Similar to the innervation of the median antenna, the main nerve of the lateral antennae splits into two main nerve roots running alongside the antenna coelom (llanr, mlanr1) in *P. bellii*, *Marphysa* spec. and *L. torquata* (Figure 7A). Prior to entering the dorsal brain neuropil, both main nerve roots split into minor nerve roots. The lateral nerve root of the lateral antenna (llanr) splits into two minor nerve roots (latla1, matla1) (Figure 7A). The more medially located nerve root of the lateral antenna (mlanr1) splits into three minor nerve roots (mlanr2, latla2, matla2) (Figure 7A). Both lateral association tracts of the lateral antenna (latla1, latla2) fuse in the dorsal brain neuropil creating the first association tract of the lateral antenna (atla1) (Figure 7A). The atla1 proceeds ventrally and fuses with the dorsal fibril mass (dfm) (Figures 7A and 8C). Both medially located association tracts of the lateral antenna (matla1, matla2) fuse in the dorsal brain neuropil, constituting the second association tract of the lateral antenna (atla2). The atla2 proceeds diagonally through the dorsal brain neuropil and fuses with the atma, creating the atmla (Figure 7A). Additionally, a further minor nerve root (mlanr3) splits from the matla2. This nerve root, the third median nerve root of the lateral antenna (mlanr3), roots in the horizontally proceeding commissure of the association system (cas) (sensu Orrhage 1995) (Figure 7A). The cas is a commissure that connects the matla2 of both lateral antennae via the mlanr3. The cas proceeds anteriorly originating in the dorsal brain neuropil.

3.4.2. Innervation Patterns in *Hyalinoecia tubicola* (Onuphidae)

The innervation pattern of the median antenna of *Hyalinoecia tubicola* (5 head appendages) concurs with that of eunicids bearing five head appendages (Figure 7). Due to the oblique cutting plane of the histological sections, tracking of association tracts was impeded in *H. tubicola*.

Lateral antennae in *H. tubicola* are innervated differently: although the lateral antenna is innervated by one main nerve (lan), too, it does not split into a pair of minor nerves. Instead, the lateral antenna nerve proceeds along the medially located side of the lateral antenna coelom towards the brain (Figure 7B). Prior to entering the dorsal brain neuropil, the lan roots at least into two minor nerve roots (lanr, atla). The more medially located nerve root (lanr) originates from the dcdr, while the association tract (atla) proceeds in a ventral direction through the dorsal brain neuropil (dnp) (Figure 7B). Before reaching the ventral part of the dorsal brain neuropil, the atla splits into two minor tracts, the median and the lateral association tract (latla, matla). The matla fuses with the respective association tract of the median antenna (atma) at the ventral side of the dorsal brain neuropil (Figure 7B). The product of this fusion is the paired association tracts of the median and the lateral antennae (atmla) (Figure 7B). The atmla run in the anterior direction through the dorsal brain neuropil and originate medially in the neuropil. The latla proceeds in the ventral direction through the dorsal brain neuropil rooting in the dorsal fibril mass (dfm) (Figure 7B).

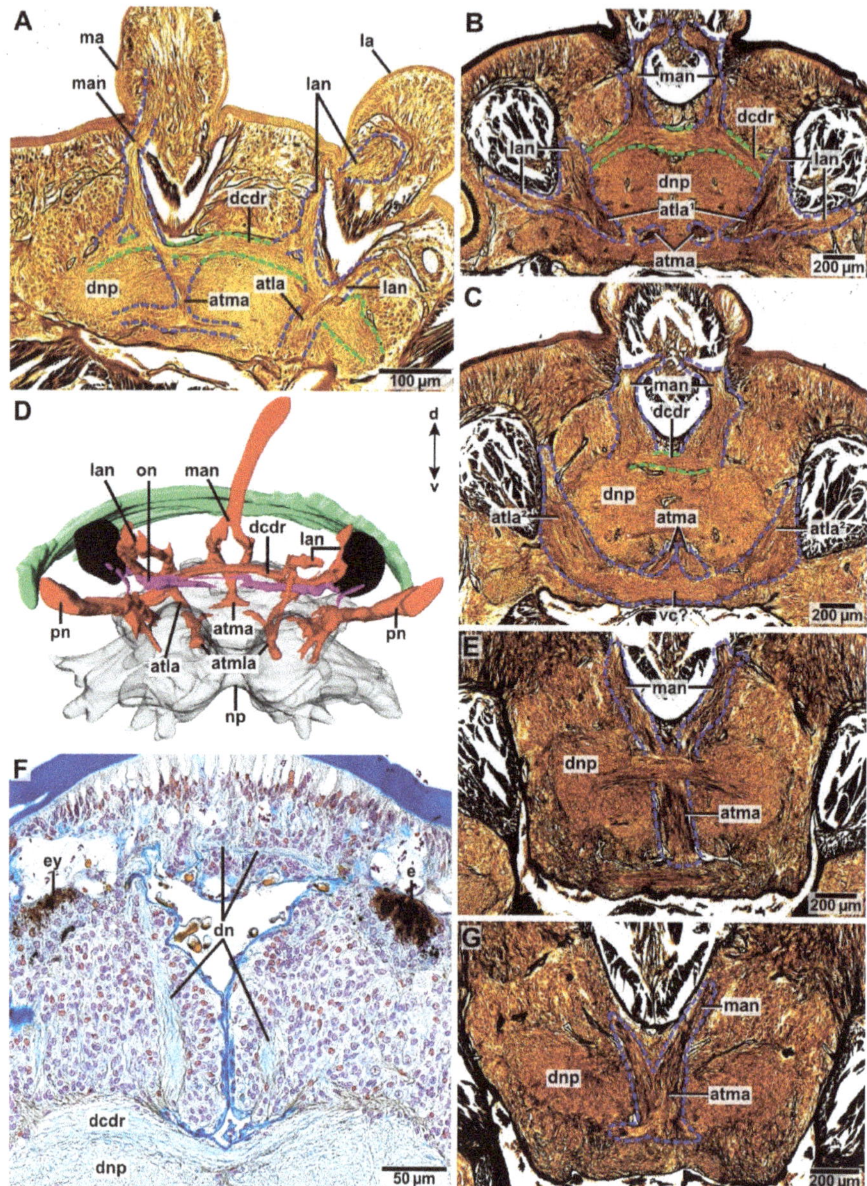

Figure 8. Association tracts of *Pauchibranchia bellii*, *Marphysa* spec., *Leodice torquata* and dorsal nerves of *Arabella iricolor*. Histological cross sections (5 µm), silver staining (**A–C,E,G**), AZAN staining (**F**). Three-dimensional reconstructions based on histological serial sections (**D**). Pictures (**B,C,E,G**) show association tracts in a consecutive anterior-posterior course. (**A**) *Marphysa* spec. The nerves of the median and the lateral antenna (man, lan) root in the dorsal commissure of the dorsal root of the circumoesophageal connective (dcdr) and the respective association tracts (atma, atla). The association tract of the median antenna (atma) proceeds in a ventral direction through the dorsal brain neuropil (dnp). The association tract of the lateral antenna (atla) runs diagonally through the dorsal brain neuropil. (**B**) *Leodice torquata*. The association tracts of the lateral antennae (atla) fuse

with the association tracts of the median antenna (atma) in the ventral part of the dorsal brain neuropil (dnp). (**C**) *L. torquata*. The association tracts of both lateral antenna (atla) fuse in the ventral brain neuropil creating a ventral commissure (vc). (**D**) *Paucibranchia bellii*, frontal view. The association tracts of the median antenna (atma) and the lateral antennae (atla) fuse in two association tracts (atmla). The association tracts of the median and the lateral antennae (atmla) proceed in the anterior direction through the dorsal brain neuropil. (**E**) *L. torquata*. Both nerves of the median antenna (man) fuse in one association tract (atma). (**F**) *Arabella iricolor*. A pair of dorsal nerves (dn) originates from the dorsal brain neuropil (dnp) in a topologically similar position as the median antenna nerves of antennae-bearing species. (**G**) *L. torquata*. The association tract of the median antenna (atma) splits into two tracts, which will fuse with the respective association tracts of the lateral antenna (atla) creating the association tracts of the median and the lateral antennae (atmla). atla: association tract of the lateral antenna; atma: association tract of the median antenna; atmla: association tract of the median and the lateral antennae; d: dorsal; dcdr: dorsal commissure of the dorsal root of the circumoesophageal connective; dn: dorsal nerves; dnp: dorsal brain neuropil; ey: eye; la: lateral antenna; lan: lateral antenna nerve; ma: median antenna; man: median antenna nerve; np: neuropil; on: optical nerve; pn: palp nerve; vc: ventral commissure; v: ventral.

3.4.3. Innervation Patterns in Eunicidae (Three and One Head Appendages)

In *Lysidice ninetta* (three antennae) and *Lysidice unicornis* (one median antenna), antennae are innervated by one main nerve (man/lan). As in *P. belli*, *Marphysa* spec. and *L. torquata*, the main nerves split at the height of the ceratophore into two nerve roots (2x manr1/llanr, mlanr1), running along the sides of the respective antenna coelom towards the dorsal brain neuropil. At least two minor nerve roots (manr2, manr3) are present in the median antenna of both species. The manr2 as well as the manr3 root in the dcdr. Distinct association tracts were not observed for the median antenna.

In *L. ninetta*, the lateral root of the lateral antennae nerve (llanr) splits into two minor nerve roots. Whether these nerve roots are association tracts is not clear, since tracking is impeded by the oblique cutting plane. At least one nerve root seems to proceed in ventral direction through the dorsal brain neuropil and could be interpreted as an association tract. As the association tracts (atla1) of *P. belli*, *Marphysa* spec. and *L. torquata*, it seems to originate at least in close proximity to the dorsal fibril mass (dfm). The other nerve root seems to originate more dorsally in the dorsal brain neuropil. The median root of the lateral antenna nerve (mlanr1) proceeds in one specimen of *L. ninetta* until the dorsal brain neuropil, but nerve roots could not be distinguished. In another specimen of the same species, the course of the mlanr1 could not be tracked or it simply shows no connection to the dorsal brain neuropil.

3.4.4. Innervation Patterns in *Dorvillea* spec. (Dorvilleidae)

In *Dorvillea* spec. (two lateral antennae, two palps) each antenna is innervated by one main nerve similar to the conditions found in the remaining species studied. Identification of antennal nerve roots is hampered by the oblique cutting plane of the histological sections. Nevertheless, two nerve roots (mlanr1, llanr) and several minor nerve roots could be identified as innervating the lateral antennae. The llanr is much more prominent than the relatively small and inconspicuous mlanr1, but both nerve roots proceed along the sides of the lateral antenna coelom towards the brain. The llanr roots at least with two minor nerve roots in the dorsal brain neuropil. The number of nerve roots of the mlanr1 as well as association tracts generally could not be identified.

3.4.5. Conditions in Oenonidae and Lumbrineridae (No Head Appendages)

Arabella iricolor (Oenonidae) lacks head appendages, but a pair of dorsal nerves (dn) is present resembling the median antenna nerves of the remaining taxa topologically (Figures 3F,H,I, 4B,C and 8F). The paired dorsal nerves surround the median coelomic cavity and root in a tract connecting the lateral neuropils (tlnp) as well as the dorsal brain neuropil (Figure 4B,C). At the insertion point of the dorsal nerves, the lateral neuropil tract forms a cushion-like neuropil area located medially in the dorsal brain neuropil (Figure 4C). This cushion-like neuropil area is surrounded by intermediate filaments of radial glia cells. Association tracts (sensu Orrhage 1995) are missing. *Scoletoma tetraura* and *Scoletoma fragilis* (Lumbrineridae) lack head appendages, association tracts and dorsal nerves (Figure 4E).

3.5. Prostomium and Prostomial Sensory Organs

In the examined representatives of Oenonidae and Lumbrineridae the prostomium is elongated and tapered (Figure 1G,H) while in Eunicidae and Onuphidae it is roundish with or without distinct buccal lips (Figure 1A–E). In Dorvilleidae the prostomium is very small and roundish (Figure 1F).

Prostomial sensory organs occur in species-specific combinations and morphology in Eunicida (Figure 1, Table 1). Antennae, palps, eyes and nuchal organ(s) are located on the posterior part of the prostomium dorsally to dorso-laterally while buccal lips are located ventrally (Figures 1 and 2). The nuchal organ is the only prostomial sensory organ completely internalized: not more than the opening(s) of the nuchal cavity may be recognized from the outside (Figure 1).

Table 1. Number of sensory prostomial structures (antennae, palps, buccal lips, eyes and nuchal organ) in the examined Eunicida.

Species	Median Antenna	Lateral Antennae	Palps	Buccal Lips	Eyes	Nuchal Organ
Pauchibranchia bellii	1	2	2	2	2	unpaired, arc-shaped
Marphysa spec.	1	2	2	2	2	unpaired, arc-shaped
Lysidice unicornis	1	0	0	2	2	unpaired, arc-shaped
Lysidine ninetta	1	2	0	2	2	unpaired, arc-shaped
Leodice torquata	1	2	2	2	2	unpaired, arc-shaped
Hyalinoecia tubicola	1	2	2	2	2	unpaired, arc-shaped
Dorvillea spec.	0	2	2	0	4	paired, simple
Arabella iricolor	0	0	0	0	4 (5)	paired, branched
Scoletoma tetraura	0	0	0	0	0	unpaired, x-shaped
Scoletoma fragilis	0	0	0	0	0	unpaired, x-shaped
Histriobdella homari	1	2	2	0?	0?	"oval, very small pits"

3.6. Head Appendages (Antennae and Palps)

Paucibranchia bellii, *Marphysa* spec., *Leodice torquata* (Eunicidae) and *Hyalinoecia tubicola* (Onuphidae) possess one median antenna, one pair of lateral antennae and one pair of palps (Figure 1A,D,E, Table 1). One pair of lateral antennae and one pair of palps are present in *Dorvillea* spec. (Dorvilleidae) (Figure 1F, Table 1), while *Lysidice ninetta* (Eunicidae) bears one median antenna and one pair of lateral antennae (Figure 1C, Table 1). One median antenna is present in *Lysidice unicornis* (Eunicidae) (Figure 1B, Table 1). *Arabella iricolor* (Oenonidae), *Scoletoma tetraura* and *Scoletoma fragilis* (Lumbrineridae) lack head appendages (Figure 1F,G, Table 1. Outer Morphology of Antennae and Palps).

Antennae and palps are digitiform and possess a basal ceratophore (antennophore/ palpophore) and a distal ceratostyle (antennal style/palpostyle) (Figure 1). Antennae and palps have a white-beige pigmented, smooth surface in *P. bellii* (Eunicidae), *Marphysa* spec.

(Eunicidae), *L. unicornis* (Eunicidae), *L. ninetta* (Eunicidae) and *Dorvillea* spec. (Dorvilleidae) (Figure 1A–C,E,F). In *L. torquata*, the ceratophores and palpophores have a white pigmentated, smooth surface, while the ceratostyles and palpostyles are annulated in white and red (Figure 1D). In *H. tubicola*, the ceratophores and palpophores are annulated, while the ceratostyles and palpostyles have smooth surfaces (Figure 1E). Further, differences in the length of antennae and palps appear between the species: *H. tubicola* and *Dorvillea* spec. possess relatively long head appendages when compared to the size of their prostomia (Figure 1D,E). *P. bellii*, *Marphysa* spec. and *L. torquata* bear head appendages of moderate length (Figure 1A,D) while antennae of *L. unicornis* and *L. ninetta* are relatively short compared to the size of their prostomia (Figure 1B,C).

Inner Morphology of Antennae and Palps

Besides a similar outer morphology, antennae and palps share a similar inner anatomy in most of the species studied. Generally, one nerve consisting of longitudinal and circular directed nerve fibres can be found in the ceratostyles as well as the palpostyles (Figure 9A,B). Longitudinal nerve fibres are located medially and are surrounded by circular nerve fibres (Figure 9A). Nuclei of glial cells can be found in both kinds of nerve fibres (Figure 9A). Both nerve fibre-types run through the ceratostyle/palpostyle until the tip of the respective head appendage (Figure 9B). A coelomic cavity (here referred to as "antenna coelom" or "palpus coelom") containing musculature is located in the ceratophores and the palpophores of all species investigated (e.g., Figure 9C–H). While the composition of antennae in *Dorvillea* spec. is the same as in the remaining taxa, its palps are differently organized: the coelomic cavity and the musculature are not limited to the palpophore, but extend until the tip of the respective head appendage (Figure 9F,G).

3.7. Innervation of Palps

In the studied Eunicidae and Onuphidae, palps are innervated by one main nerve (pn) running along the more medially located side of the palp coelom towards the brain (Figure 9D,E). Before entering the brain, the palp nerve splits into several minor nerve roots—in most species studied three minor palp nerve roots (pnr1, pnr2, pnr3) are present (Figure 7). The two more dorsally located palp nerve roots (pnr1, pnr2) originate from the dorsal brain neuropil (Figures 7 and 9E). The more ventrally located palp nerve root originates from the ventral brain neuropil (Figures 7E and 9E). The exact number of minor nerve roots was in some specimens hard to identify because oblique cutting planes hampered tracking.

In *Pauchibranchia bellii*, we found 3–4 minor nerve palp roots (3 rooting in the dnp (one of which rooting in the dfm) and probably 1 rooting in the vnp). For *Marphysa* spec. we could identify only one minor palp nerve root, rooting in the dfm (dorsal brain neuropil). Probably more nerve roots are present. In *Leodice torquata*, up to four or more minor palp nerve roots can be present depending on the specimen we are looking at. There were 1–2 nerve roots originating from the vnp, while 2–3 nerve roots were found rooting in the dnp. The exact number of palp nerve roots for *Hyalinoecia tubicola* could not be determined due to the oblique cutting plane.

In *Dorvillea* spec. two distinct palp nerves (dpn, vpn) innervate each palp (Figure 9F,G). Both palp nerves proceed from the basal to the distal end of each palp running along both sides of the coelomic cavity. The more ventrally located palp nerve (vpn) fuses via at least two minor nerve roots with the ventral brain neuropil, while the more dorsally located palp nerve (dpn) and its minor nerve roots originate from the dorsal brain neuropil. At least two minor nerve roots were observed for np.

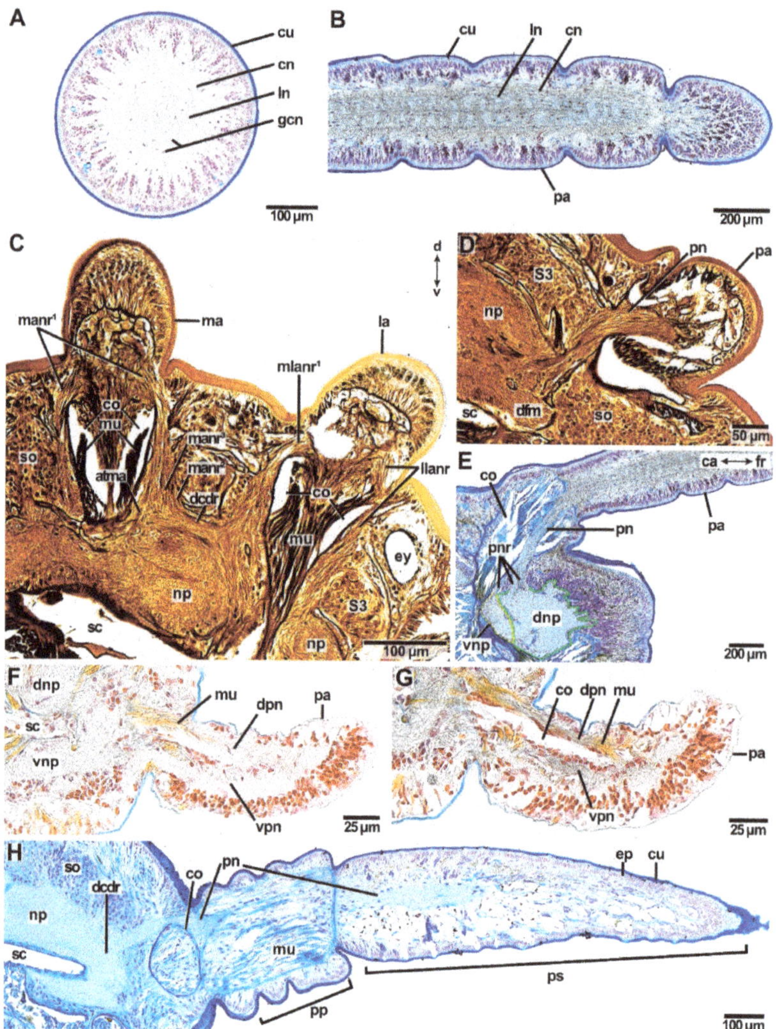

Figure 9. Inner morphology and innervation of antennae and palps. Histological cross sections (5 μm) stained in AZAN (**A**,**E–H**) or silver (**C**,**D**). Horizontal section (5 μm, AZAN) (**B**). (**A**) *Leodice torquata* (lateral antenna). Antennae as well as palps possess one main nerve, which consists of medially located longitudinal nerve fibres (ln) and surrounding circular nerve fibres (cn). Both types of nerve fibres possess nuclei of glia cells (gcn). (**B**) *Leodice torquata* (palp). Longitudinal nerve fibres (ln) and circular nerve fibres (cn) proceed nearly until the tip of the respective head appendage. (**C**) *Marphysa* spec.: The main nerve of the median antenna and the lateral antenna splits at height of the coelomic cavity (co) into two main roots (median antenna: 2x manr1; lateral antenna: mlanr1, llanr), respectively. Each main nerve root runs alongside the antenna coelom (palp coelom) (co) towards the brain (np). Prior to entering the brain, the main nerve roots split into several minor nerve roots (e.g., manr2, manr3, atma), some of which originate from the dcdr or proceed as association tracts (e.g., atma) through the dorsal brain neuropil. Each antenna or palp coelom (co) contains musculature (mu). (**D**) *Marphysa* spec.: The palp nerve (pn) proceeds along the dorsally located side of the palp coelom (co) and splits into minor nerve roots prior to entering the brain (np). (**E**) Palp nerve roots exemplified for *Leodice torquata*. The palp nerve (pn) passes the palp coelom (co) and

splits at least into three minor palp nerve roots (pnr) before entering the brain. The two more dorsally located palp nerve roots originate from the dorsal brain neuropil (dnp), and the more ventrally located palp nerve root originates from the ventral brain neuropil (vnp). (**F,G**) In *Dorvillea* spec. each palp is innervated by two palp nerves (dpn, vnp). The more dorsally located palp nerve (dpn) originates from the dorsal brain neuropil (dnp), and the more ventrally located palp nerve (vpn) originates from the ventral brain neuropil (vnp). Both nerves proceed together with the coelomic cavity (co) and musculature (mu) through the palp. (**H**) Separation of palps into palpophore (pp) and palpostyle (ps) exemplified for *Hyalinoecia tubicola*. The palp is composed of an annulated basal palpophore (pp) and a distal palpostyle (ps). One palp nerve (pn) innervates the palp. atma = association tract of the median antenna, ca = caudal, cn = circular nerve fibres, co = coelom, cu = cuticle, d = dorsal, dcdr = dorsal commissure of the dorsal root of the circumoesophageal connective, dfm = dorsal fibril mass, dnp = dorsal brain neuropil, dpn = dorsal palp nerve, ep = epidermis, ey = eye, fr = frontal, gcn = nuclei of glia cells, la = lateral antenna, lan = lateral antenna nerve, llanr = lateral root of the lateral antenna nerve, ln = longitudinal nerve fibres, ma = median antenna, manr1 = first root of the median antenna nerve, manr2 = second root of the median antenna nerve, manr3 = third root of the median antenna nerve, mlanr1 = median root of the lateral antenna nerve, mu = musculature, np = neuropil, p = palp, pn = palp nerve, pnr = roots of palp nerve, pp = palpophore, ps = palpostyle, S3 = somata of class 3 neurons, sc = stomatogastric cavity, so = somata, v = ventral, vnp = ventral brain neuropil, vpn = ventral palp nerve.

3.8. Buccal Lips

Distinct buccal lips occur as paired structures in front of the mouth opening on the ventral side of the prostomium in the studied specimens of Eunicidae and *Hyalinoecia tubicola* (Onuphidae) (Figure 10A,B). In Eunicidae one pair of more or less rectangular-shaped buccal lips is present with a smooth surface (Figure 10A). In *H. tubicola* two pairs of buccal lips are present—one dorsal (dbl) and one ventral (vbl) pair –plus on additional pair of accessory buccal lips (abl) (Figure 10B). The more or less rectangular-shaped ventral buccal lips of *H. tubicola* are placed directly in front of the mouth opening and have a smooth surface. The dorsal buccal lips and the accessory buccal lips are spindle-shaped with a rough surface and are positioned in pairs anteriorly to the ventral buccal lips. Buccal lips are innervated by nerves (bln) originating from the ventral brain neuropil (Figure 10D,E). The buccal lip nerves proceed in the ventral direction alongside the buccal cavity (blc) and fuse eventually, building a loop. All remaining taxa studied belonging to the families of Dorvilleidae, Oenonidae and Lumbrineridae lack buccal lips (exemplified for Oenonidae, Figure 10C).

3.9. Eyes

Eyes are multicellular cerebral eyes of the pigment cup type and can be present as two pairs, one pair, or they are absent (Figures 1 and 11, Table 1). In species possessing one pair of eyes (*Paucibranchia bellii*, *Marphysa* spec., *Lysidice unicornis*, *Lysidice ninetta*, *Leodice torquata*; *Hyalinoecia tubicola*), eyes are located between the lateral antennae and the palps (Figure 1A,D,E), or in a topologically equal position, if head appendages are lacking (Figure 1B,C,G). In species possessing two pairs of eyes (*Dorvillea* spec.; *Arabella iricolor*), the more anteriorly located pair of eyes resemble those of species with one pair of eyes in topology (Figure 11). The posteriorly located pair of eyes is smaller than the anterior pair of eyes (Figure 11B,E). In *Dorvillea* spec. the posterior pair of eyes is located dorsally to the lateral antennae (Figure 1F), while the posterior pair of eyes of *A. iricolor* can be found in an equivalent position (Figure 1G). In some of the studied specimens (*A. iricolor*, *L. torquata*) additional small eyes of the same type can be found next to the "regular" eyes (Figure 11B).

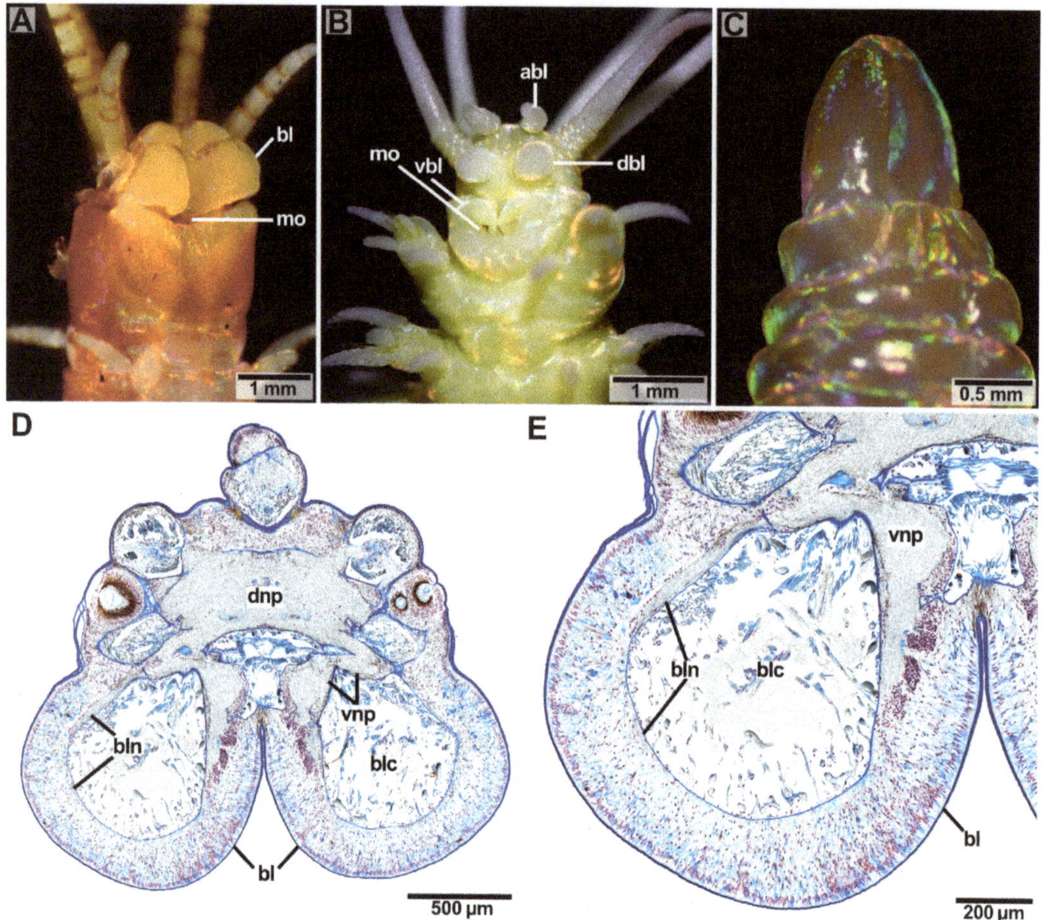

Figure 10. Anatomy and innervation of buccal lips. Ventral view of the prostomium (**A,B**). Histological cross sections (5 μm, AZAN) (**D,E**). (**A**) Paired buccal lips (bl) of *Leodice torquata*. (**B**) In *Hyalinoecia tubicola* two pairs of buccal lips are present—one pair of dorsal buccal lips (dbl) and one pair of ventral buccal lips. Additionally, one pair of accessory buccal lips (abl) are present. (**C**) In *Arabella iricolor* buccal lips are absent. (**D,E**) Innervation of buccal lips (bl) in *L. torquata*. Buccal lips (bl) are innervated by nerves (bln) originating from the ventral brain neuropil (vnp). The nerves run along the cavity of the buccal lip (blc). abl = accessory buccal lips, bl = buccal lip, blc = coelomic cavity of the buccal lip, bln = buccal lip nerve, dbl = dorsal buccal lip, dnp = dorsal brain neuropil, mo = mouth opening, vbl = ventral buccal lip, vnp = ventral brain neuropil.

3.9.1. Inner Morphology of Eyes

The eyes of the examined Eunicida share the same inner structure. The microvilli of the photoreceptor cells project into the eye cup, which is made up of pigmented supportive cells (Figure 11A). The layer of pigmented supportive cells is surrounded by a dense layer of photoreceptor cells (Figure 11A). In *Marphysa* spec., *L. ninetta*, *L. torquata*, *H. tubicola* and *Dorvillea* spec. a cuticular invagination projects from the outer cuticle of the animal into the eye (Figure 11A,B,C). In all these species—except for *Marphysa* spec.—this invagination forms a lens-like structure inside the eye.

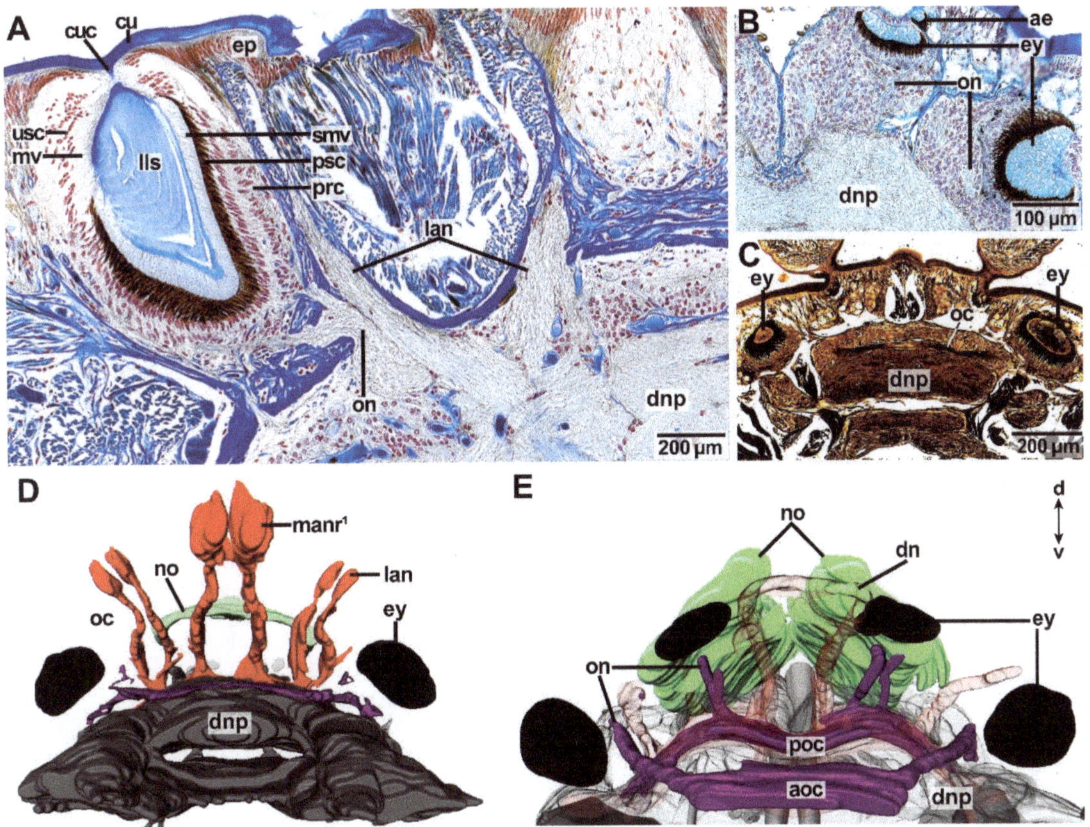

Figure 11. Anatomy and innervation of the eyes. Histological cross sections (5 μm), AZAN staining (**A**,**B**), silver staining (**C**), 3D reconstructions (**D**,**E**). (**A**) *Paucibranchia bellii*. The eye is composed of a lens-like structure (lls) in the inner, covered by a layer of sensory microvilli (smv), pigmented supportive cells (psc), and photoreceptor cells (prc). Unpigmented sensory cells (usc) and microvilli (mv) cover the lens-like structure as the uppermost layer. The eye is situated basally to the epidermis (ep). A cuticular channel (cuc) can be identified projecting from the cuticle (cu) into the eye. The eye is innervated by an optical nerve (on) originating from the brain (np). The optical nerve (oc) splits next to the eye into minor nerves. (**B**) *Arabella iricolor*. Two pairs of eyes are present, the ventral one being bigger than the dorsal one. Both eye pairs are innervated by one optical nerve (on), respectively. (**C**,**D**) *Lysidice ninetta*. In the brain (np), the optical nerves are connected by an optical commissure (oc). (**E**) *Arabella iricolor*. Two optical commissures can be differentiated in the brain (np)—an anterior optical commissure (aoc) connects the optical nerves of the more ventrally located pair of eyes and a posterior optical commissure (poc) connects the optical nerves of the more dorsally located pair of eyes. ae = additional eye, aoc = anterior optical commissure, cu = cuticle, cuc = cuticular channel, dn = dorsal nerve, ep = epidermis, ey = eye, lan = lateral antenna nerve, lls = lens-like structure, man = median antenna nerve, mv = microvilli, no = nuchal organ, np = neuropil, on = optical nerve, poc = posterior optical commissure, prc = photoreceptor cell, psc = pigmented sensory cell, smv = sensory microvilli, usc = unpigmented sensory cells.

3.9.2. Innervation of Eyes

Each eye is innervated by one distinct optical nerve, which splits into fine nerves next to the eye (Figure 11A,B,E). The optical nerves fuse in the brain by forming an optical commissure (Figure 11C–E). In species possessing one pair of eyes, one optical commissure is present (Figure 11C,D), in species possessing two pairs of eyes two optical commissures are present (Figure 11E). The optical commissure connecting the more anteriorly located pair of eyes is positioned anteriorly to the second optical commissure as well as the dorsal commissure of the dorsal root of the circumoesophageal connectives (dcdr) (Figure 11E).

3.10. Nuchal Organs

All species studied possess a nuchal organ. The nuchal organ invaginates from the dorsal surface right behind the median antenna—if present—building either an unpaired or a paired ciliated cavity, the latter having separate openings. The shape of the nuchal organ can vary between families. The unpaired nuchal organ is either arc- or x-shaped (Figure 12A–C,F–I,L, Table 1), while the paired nuchal organ can be a simple invagination or an elaborate branching structure (Figure 12D,E,J,K, Table 1). All nuchal organs end blindly.

3.10.1. Morphology of Nuchal Organs

Nuchal organs share a similar composition among the species studied. A relatively thick layer of cuticle lines the nuchal cavity (Figure 12A–F) and patches of cilia extend into the nuchal cavity perforating the cuticle (Figure 12B,D). Two patches of cilia can be found in the paired as well as in the unpaired nuchal organ. In the arc-shaped nuchal organs, these ciliary patches are located on the ventral side of the nuchal organ (Figure 12A,B,H,I). In the branched and x-shaped nuchal organs, the ciliary patches are located in the dorsal branches of the nuchal cavity (Figure 12E,F,K,L). In all nuchal organs, the ciliary patches originate at the lateral sides of the nuchal cavity. More posteriorly, the ciliary patches enlarge until the basal side of the nuchal organ is covered nearly completely with cilia (Figure 12B). Only a small area in the middle of the basal side is not covered with cilia (Figure 12B). The ciliary patches can be everted into the nuchal cavity (Figure 12B).

3.10.2. Innervation of Nuchal Organs

Innervation of the ciliary patch cells is provided by fine nerves present in the tissue ventrally to the nuchal organ (Figure 12B,E–G). These fine nerves bundle in the paired nuchal organ nerves, which originate from the posterior-most part of the dorsal brain neuropil (Figures 2G and 12E,F,M,O). No nuchal commissure was observed in the brain neuropil. Adjacent to the nuchal organ nerve, intermediate filaments of glial cells are present. Longitudinal and transversal musculature attaches at the lateral ends of the nuchal organ proceeding in the ventral direction until reaching the oesophagus. The nuchal cavity ends approximately at the height of the oesophagus and only musculature attaching posteriorly to the nuchal organ is still present. The ciliated patches of *L. unicornis* are covered by an orange-coloured layer in the AZAN-stained sections—possibly a secretion of gland cells.

3.10.3. Shape Differences of the Nuchal Organ

Regarding the shape of the nuchal organ, the following clusters can be found. The examined specimen of Eunicidae and Onuphidae possess an unpaired, arc-shaped nuchal organ (Figure 12A–C,H,I). The nuchal organ of *Dorvillea* spec. is a paired, simple invaginated structure (Figure 12D,J), *A. iricolor* possesses a paired, branched nuchal organ (Figure 12E,K,M,N) and the examined representatives of Lumbrineridae possess an unpaired, x-shaped nuchal organ (Figure 12F,L,O, Table 1).

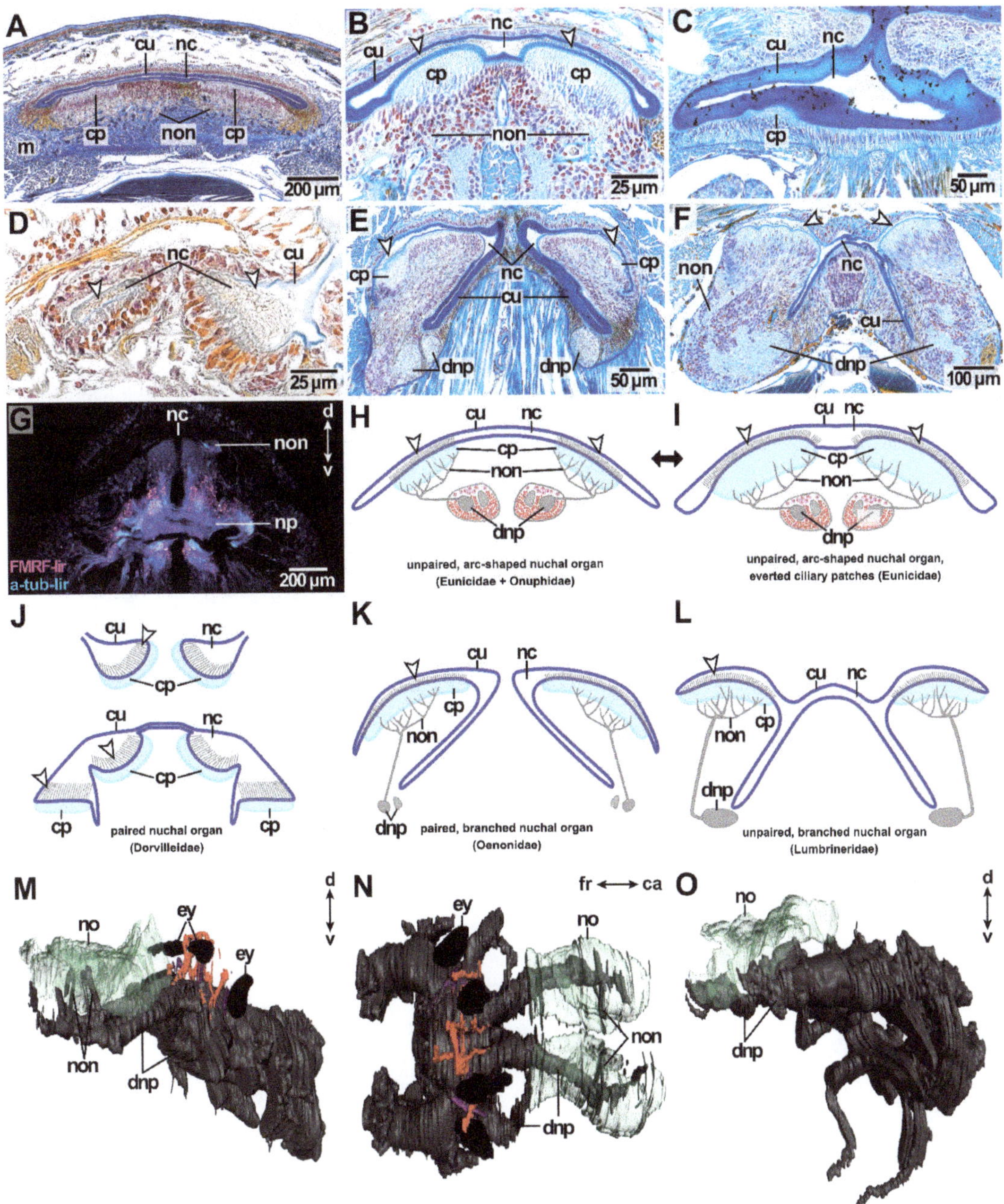

Figure 12. Nuchal organs. Histological cross sections (5 µm) stained in AZAN (**A–G**) and immuno-histochemically stained section (60 µm) (**H**). Schematic drawings (**I–M**), 3D reconstructions based on histological sections (**N–P**). (**A**) *Paucibranchia bellii*, unpaired, arc-shaped nuchal organ. (**B**) *Lysidice ninetta*, unpaired, arc-shaped nuchal organ. (**C**) *Hyalinoecia tubicola*, unpaired, arc-shaped nuchal organ. (**D**) *Dorvillea* spec., paired nuchal organs. (**E**) *Arabella iricolor*, paired, branched nuchal organ. (**F**) *Scoletoma tetraura*, unpaired, x-shaped nuchal organ. (**G**) *L. ninetta*, nuchal organ nerves (non)

emanating from the brain (np) and innervating the nuchal organ. (**H,I**) Unpaired, arc-shaped nuchal organ in Eunicidae and Onuphidae. (**J**) Paired nuchal organ of *Dorvillea* spec. (**K**) Paired, branched nuchal organ of Oenonidae. (**L**) Unpaired, branched nuchal organ of Lumbrineridae. (**M**) Topology of the paired, branched nuchal organ in *A. iricolor*, lateral view. The nuchal organ (no) is located dorsally to dorsal brain neuropil (dnp) and is innervated by nerves (non) emanating from the posterior part of the brain. (**N**) Topology of the paired, branched nuchal organ in *A. iricolor*, dorsal view. (**O**) Topology of the unpaired, branched nuchal organ in *S. tetraura*, lateral view. The nuchal organ (no) is located dorsally to the posterior part of the dorsal brain neuropil (dnp). White arrow = cilia. ca = caudal, cu = cuticle, cp = ciliary patch, d = dorsal, dnp = dorsal brain neuropil, ey = eye, fr = frontal, m = musculature, nc = nuchal cavity, no = nuchal organ, non = nuchal organ nerve, np = neuropil, v = ventral.

Some peculiarities occur in the nuchal organs of Dorvilleidae, Oenonidae and Lumbrineridae. First of all, in *Dorvillea* spec. the nuchal cavities are interconnected anteriorly and they are connected to ciliary patches located on the lateral sides of the prostomium. Two of these ciliary patches are located on both sides of an elevation between the lateral antenna and the palp. More anteriorly, the elevation shrinks and the two ciliated patches fuse. In the posterior course, the more dorsally located ciliated patch is connected to the nuchal organ of the respective side (Figure 12J).

In *A. iricolor* (Oenonidae), the nuchal cavities start as simple invaginations resembling the arc-shaped condition of Eunicidae and Onuphidae but being separated. In the posterior course, the cavities branch due to lateral eversions. The shape of the nuchal organs resembles facing arrow heads (">" "<") at this point (Figure 12E,K). More posteriorly, the dorsal and ventral branches split eventually, so that four branches are present ending blindly.

Representatives of Lumbrineridae (*S. fragilis*, *S. tetraura*) possess a single-grooved, x-shaped nuchal organ (Figure 12F, Table 1). Similar to *A. iricolor*, the nuchal organ resembles the arc-shaped condition only shortly after its invagination. At this point, the lateral ends of the nuchal organ are enlarged building ciliated cavities already differing from the condition in Eunicidae and Onuphidae. These enlarged chambers are covered by cilia, while the remaining parts are covered by cuticles only. More posteriorly, the nuchal organ branches and resembles an "x" in its shape (Figure 12F,L). The branches are still connected to each other contrary to the condition in *A. iricolor* (Oenonidae). More posteriorly, the nuchal organ is separated in the middle and ends blindly.

4. Discussion

4.1. Phylogenetic Significance of Sensory Organs in Eunicida

Sensory organs in Eunicida are quite diverse and show variations in number, shape, and innervation. Therefore, it is possible to group certain characters (Table 2) and draw homology hypotheses about their evolutionary origin.

A mapping of characters on the latest phylogenetic tree of Eunicida (based on Tilic et al. [2]) delivers the following hypotheses (Figure 13):

1. Antennae, palps, two pairs of multicellular adult eyes, nuchal organs and lateral organs seem to have already evolved in the common stem lineage of Eunicida. It is most parsimonious to assume, that the stem species of Eunicida possessed two pairs of multicellular adult eyes. In this scenario, a reduction of adult eyes took place independently in Lumbrineridae (complete loss of adult eyes), Eunicoidae (reduction to 1 pair of adult eyes) and Histriobdellidae (complete loss of adult eyes?). In the literature, there is no report about eyes in Histriobdellidae, but according to Rouse et al. [1], eyes may be present in *Steineridrilus cirolanae* (Führ, 1971). It can be assumed, that the stem species of Eunicida possessed a roundish prostomium and that in Oenonoidea a change of shape to a tapered, elongated prostomium occurred (Figure 13). This shift concurs with the hypothesized change of lifestyle from epi- to endobenthic.

Table 2. Character matrix. Depicted is the distribution of characters among the six studied families of Eunicida. The presence of a character is indicated by "+", while its absence (or reduction) is marked with "-". abl: accessory buccal lips, at: association tracts, dbl: dorsal buccal lips, ey: eye, la: lateral antenna, lo: lateral organ, ma: median antenna, no: nuchal organ, pa: palp, pr: prostomium, vbl: ventral buccal lips.

Tree Code	Character	Lumbrineridae	Oenonidae	Onuphidae	Eunicidae	Dorvilleidae	Histriobdellidae
1	ma	-	-	+	+	-	+
2	la	-	-	+	+	+	+
3	pa	-	-	+	+	+	+
4	vbl	-	-	+	+	-	-
5	dbl	-	-	+	-	-	-
6	abl	-	-	+	-	-	-
7	ey	-	+ 2 pairs	+ 1 pair	+ 1 pair	+ 2 pairs	-
8	no	+	+	+	+	+ 2 pairs	+
9	shape no	unpaired, x-shaped	paired, branched	unpaired, arc-shaped	unpaired, arc-shaped	paired, simple	paired, simple
10	lo	+	+	+	+	+	+
11	at	-	-	+	+	-	?
12	shape pr	elongated, tapered	elongated, tapered	roundish	roundish	roundish	roundish

2. Antennae were reduced several times independently in Eunicida. A reduction of antennae took place once in Oenonoidea (sensu Tilic et al. [2]) with Lumbrineridae lacking antennae completely and Oenonidae having, at least in some species, remnants of either antennae, e.g., in *Oenone* (Lamarck, 1818), or antennae nerves (dorsal nerves), e.g., in *Arabella iricolor*. Dorsal nerves were observed by Zanol [20] in *Schistomeringos pectinata* (Perkins, 1979) (Dorvilleidae) and based on the similar topology and direction of nerves, she suggested a homology of dorsal nerves and median antenna nerves. Our results are in accordance with this hypothesis. Generally, it is assumed that nerves are attracted by the morphological structure (e.g., antennae) instead of initiating the formation of the structure itself [32] making a reduction of antennae even more plausible. Further, the median antenna was reduced completely in Dorvilleidae and in some specimens of Eunicidae the lateral antennae were reduced, e.g., in *Nematonereis unicornis*.

3. Palps are homologous structures among Eunicida. A complete reduction of palps took place in Oenonoidea, supporting the sister group relationship of Lumbrineridae and Oenonidae. All other eunicidan families possess palps in the ground pattern. In some specimens of Eunicida palps have been reduced, e.g., in *Lysidice ninetta* and *Nematonereis unicornis*. Similar to the morphology of eunicidan antennae, the palp coelom as well as palp musculature are limited to the palpophore in most Eunicida. This is an unusual condition when compared to palps of other annelids. The palp coelom as well as the palp musculature are normally not limited to the palpophore but extend all the way up until the tip of the palp. The only eunicidan palps observed matching this pattern are those of *Dorvillea* spec.

4. Buccal lips are not homologous to palps due to differing innervation patterns. The innervation of buccal lips was described by Orrhage [14] for *Hyalinoecia tubicola*: The dorsal buccal lips are innervated by nerves emanating from the lateral parts of the first ventral commissure of the ventral root of the circumoesophageal connective (vcvr1) while the ventral buccal lips are innervated by nerves rooting laterally in the third ventral commissure of the ventral root of the circumoesophageal connective (vcvr3).

5. Buccal lips are an autapomorphy of Eunicoidea (sensu Tilic et al. [2]). In Eunicidae solely one pair of ventral buccal lips is present, while in Onuphidae additionally, one pair of dorsal buccal lips is present. In *Hyalinoecia tubicola* one pair of accessory buccal lips was observed besides the paired dorsal and ventral buccal lips.

6. The shape of the nuchal organ can be used to infer phylogenetic relationships in the studied Eunicida. Oenonoidea have a branched nuchal organ while the nuchal organ of Eunicoidea is unpaired and arc-shaped. In Dorvilleidae nuchal organs were duplicated. In Histriobdellidae the nuchal organ is a paired, simple structure. The shape of the nuchal organ in the eunicidan ground pattern remains unresolved.

7. Association tracts can be regarded as an autapomorphy of Eunicoidea as long not proven to be present in Histriobdellidae, too. Association tracts seem to be restricted to species possessing three antennae and one pair of palps (=species belonging to Eunicoidea), making it theoretically possible that association tracts might also be present in Histriobdellidae (=possessing three antennae + one pair of palps). However, in the histological serial section of *Histriobdella homari* we could not observe any association tracts.

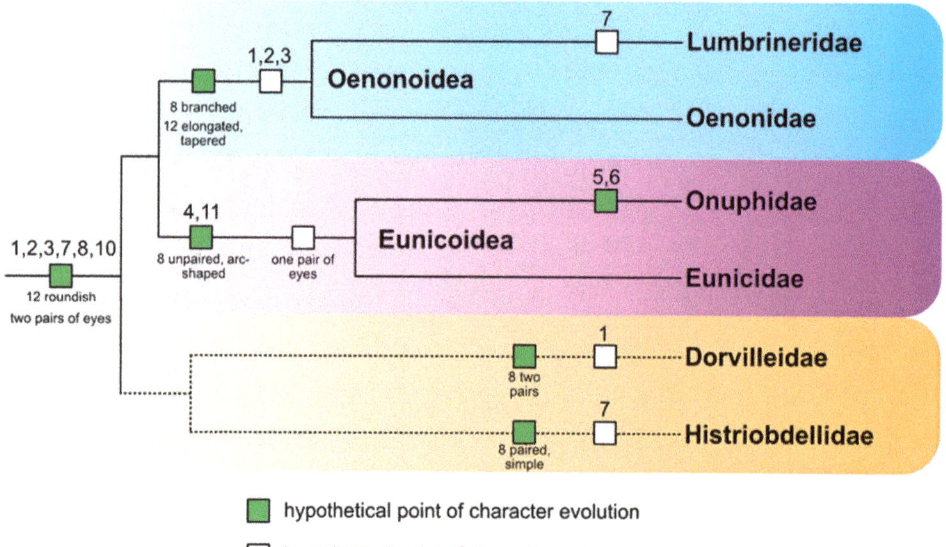

Figure 13. Characters mapped on the latest phylogenetic tree of Eunicidae (Tilic et al. [2]). Dotted lines indicate unresolved phylogenetic positions: either Dorvilleidae and Histriobdellidae are a clade and sister group to the remaining Eunicidae, or Histriobdellidae is a sister taxon to all remaining Eunicidae. Numbers code for characters listed in the character matrix (see Table 2).

4.2. Is the Brain shape Influenced by Sensory Structures?

The central nervous system of all examined Eunicida concurs with descriptions of the general annelid neuroanatomy [10,13,15,29,32,37]. Differences occur only in the condensed oesophageal ganglion of *Arabella iricolor* (Oenonidae): here, the oesophageal ganglion is much more prominent than in all other species investigated. It is additionally connected via commissures to the circumoesophageal connectives. Due to the condensed oesophageal ganglion, the brain of *A. iricolor* does not form a similar-shaped ring around the stomatogastric cavity as in all other examined species.

The brain shows a consistent morphology among the species studied: A dorsal and ventral brain neuropil can be distinguished topologically, separated by an intracerebral cavity [26,29]. As in other polychaetes, the brain can be defined as the frontal-most mass of neuropil surrounded by somata and can be interpreted as a "single entity" as already Malaquin [38] stated in his argumentation for an unsegmented brain [14]. The general brain anatomy is not affected to a great extent by sensory structures. Deviations occur only on a microanatomical level in the presence of sensory commissures, association tracts and lateral neuropils.

4.2.1. Optical Commissure

Species bearing one pair of eyes possess one optical commissure (Eunicidae, Onuphidae). Species with two pairs of eyes possess two optical commissures (Oenonidae, Dorvilleidae)—an anteriorly located one (aoc) and a posteriorly located one (poc). In *Dorvillea* spec. only the insertion points of the optical nerves could be observed and not the complete commissure due to an oblique cutting plane. The presence of an optical commissure was reported by Zanol [29] for Dorvilleidae only. The optical commissures are located dorsally in the brain and anteriorly to the dorsal commissure of the dorsal root of the circumoesophageal connective (dcdr) in all species investigated, which concurs with the findings of Orrhage and Müller [13].

4.2.2. Nuchal Commissure

A commissure of the nuchal organ nerves as Orrhage and Müller [13] and Zanol [29] reported, was not observed in any eunicidan specimen examined. The nuchal organ is innervated by nerves originating from the posterior-most part of the dorsal brain neuropil as typical for other annelids [13,29,39,40].

4.2.3. Association Tracts

Association tracts were found in species bearing five head appendages (Eunicidae, Onuphidae), while in species lacking head appendages (Oenonidae, Lumbrineridae) association tracts are not present. Interestingly, we did not observe any association tracts in Dorvilleidae (four head appendages). Together with the fact, that further studies on association tracts in Dorvilleidae are lacking, it could be assumed that association tracts might be an autapomorphy of Eunicoidea. Binard and Jeener [26] and Zanol [29] for example reported association systems (sensu Orrhage [14]) for Eunicidae and Onuphidae only—in Dorvilleidae, Zanol [29] did not find an association system. The association system is an anteriorly located structure composed of small cells [29] connected to association tracts. So, the association system seems to be involved in the processing of sensory stimuli transmitted by association tracts and in a further sense by head appendages. In our study, we observed anteriorly located areas of circularly arranged neurite bundles (atmla), in which the association tracts of the median and the lateral antennae anastomose. These areas might be association tracts on their own. If these areas concur with the association system sensu Orrhage [14] is a matter of debate—e.g., we did not observe a composition of small cells. At least in topology and the connection to association tracts both structures concur—presumably also in function as processors of sensory stimuli received by head appendages. We found the atmla only in species of Eunicoidea with five head appendages. Species with less than five head appendages belonging to Eunicidae (*Lysidice unicornis*, *Lysidice ninetta*) show neither a distinct atmla nor distinct association tracts in the histological sections. However, in *L. ninetta* one possible association tract was observed (one of the minor nerve roots of the lateral root of the lateral antennae nerve). Whether this observation consolidates a general difference in microanatomical processing of sensory stimuli received by head appendages between Eunicoidea and Dorvilleidae is a matter of debate. Clearly stated can be, that innervation patterns are more complex in species bearing five head appendages (Eunicoidea).

The number of association tracts found among the different species in this study is mostly in accordance with the pertinent literature [14]: In Eunicoidea one pair of association tracts is connected to the median antenna and in Onuphidae one pair of association tracts is linked to the lateral antennae, respectively. For the lateral antennae of Eunicidae, we found—contrary to the report of Orrhage [14] of one pair of association tracts—two pairs of association tracts. The additional lateral association tracts seem to anastomose with the ventrally located commissure (vc) in the dorsal brain neuropil of Eunicidae, resembling the dorsal fibril mass (dfm) of the schematic drawing of Orrhage [14] of *H. tubicola*. The other pair of lateral association tracts fuse with the pair of median association tracts, more or less similar to the drawing of Orrhage [14].

The presence of palps does not affect brain anatomy in Eunicida: no special brain compartments, commissures or tracts associated with palps were observable inside the brain neuropil. Sensory information gained by palps seems to be processed somewhere in the drcc and the vrcc—the commissures the palp nerves originate from. In contrast to conditions found in Polynoida and Nereididae [12,41–43], where each palp nerve is associated with mushroom body neuropil, palp nerves are not associated with mushroom bodies in Eunicida.

4.2.4. Lateral Neuropils

Another characteristic shared by the studied representatives of Oenonoidea is the paired lateral neuropils, which are lacking in Eunicoidea and Dorvilleidae. The lateral neuropils may further support the clade Oenonoidea. The lateral neuropils are interconnected via a tract in the brain neuropil, and in *A. iricolor*, the dorsal nerves originate from this tract. Further neuronal connections revealing the function of the lateral neuropils were not observed. A pair of lateral neuropils (called "dorso-ventral neuropil") anteriorly in the brain of Eunicidae, Dorvilleidae, Oenonidae and Lumbrineridae (except for Onuphidae) was also observed by Zanol [29]. This structure seems not to match the neuropils we observed in Oenonidae and Lumbrineridae, since Zanol [29] described these neuropils as connecting the dorsal and ventral neuropils via several connections (dorsal fibril mass, dorsal commissure of the ventral root of the circumoesophageal connective, first ventral commissure of the ventral root of the circumoesophageal connective). Neither we observed lateral neuropils in other taxa than Oenonidae and Lumbrineridae nor we saw it connecting dorsal and ventral neuropil via the given connections. However, the lateral neuropils as well as Zanol's dorso-ventral neuropils resemble polychaete mushroom bodies partially in shape and composition. Already Hanström [12] compared the small and chromatin-rich "Ganglienzellen" occurring in high amounts in the anterior region of the brains of *Eunice norvegica* (Linnaeus, 1767), *Hyalinoecia tubicola* and *Diopatra cuprea* (Bosc, 1802), with the cells of the corpora pedunculata. The presence of mushroom body-like structures in polychaete taxa was discussed already in the former literature [12,34,43] and it was pointed out, that mushroom body-like structures may be present in Eunicida [12]. However, the current literature (e.g., [1]) states that mushroom bodies are not present in Eunicida referring to the study of Bullock and Horridge [44]. The mushroom body-like structures in Eunicida might be remnants of mushroom bodies if a reduction is hypothesized. Since Phyllodocida, e.g., *Nereis diversicolor*, possess distinct mushroom bodies [42,43], a secondary reduction of mushroom bodies in Eunicida is not unlikely.

Regarding the presence of prominent neuropil areas such as the dorso-ventral neuropil and the lateral neuropils of Oenonidae and Lumbrineridae, the question arises, if these neuropils serve as compensators for lacking senses, e.g., of antennae, palps, eyes etc. Studies on this topic were conducted by von Haffner in 1962 [11]. His studies are based on a topographical tripartition of the brain using sensory organs as markers for the different areas (fore-brain = "palps" (=nowadays buccal lips), mid-brain = antennae and eyes, hind-brain = nuchal organ). Von Haffner [11] compared the size of the brain areas among Eunicida and proposed, that the extended hind-brain is associated with the extended nuchal organs in Oenonidae and Lumbrineridae. He deduced that the extension of both structures compensates for the missing sensory input of antennae and palps. Although we clearly saw shape differences and a generally bigger surface of the nuchal cavity in Oenonoidea when compared to the remaining Eunicida, we did not observe an extended posterior brain area (=hind-brain) in these taxa. Since the brain appeared as a continuous mass of neuropil in our study, we did not separate the brain according to Racovitza [45]. The only differentiation we applied was the distinction between a dorsal and a ventral neuropil. Both neuropils have equal sizes among the different species studied regardless of the number of prostomial sensory organs. Together with the fact, that we did not observe tracts connecting the nuchal organs with these neuropils, there is no reason to assume a compensational function of both neuropils in this aspect.

According to Zanol [29], shape differences in the brain occur among the families of Eunicida—especially in Onuphidae. She reports a different topology of the intracerebral cavity in Onuphidae (placed between the vcvr1 + vcvr2 and the vcvr3 + vcvr2) when compared to Eunicidae, Dorvilleidae, Oenonidae and Lumbrineridae (placed between the dcvr and the vcvr1). Furthermore, she observed an x-shaped connection in Onuphidae (between the vcvr1 + vcvr2 and the dcvr), which is lacking in the other families. Since our study focussed methodically on paraffin histology, the view we had on the brain differs from that of Zanol's immunohistochemical studies, e.g., we did not clearly see all commissures of the roots of the circumoesophageal connectives and with this, we did not see different topologies of the intracerebral cavity or the x-shaped connection. Further, we did not recognize shape differences in the brains of Onuphidae and Eunicidae, but it must be considered, that we studied one adult specimen of Onuphidae (*Hyalinoecia tubicola*). Zanol [29] studied *Onuphis* (*Nothria*) *iridescens* (Johnson, 1901) for ventral dissection and sagittal cuts, and juveniles of *Kinbergonuphis simoni* (Santos, Day and Rice, 1981), so that species-specific differences or age effects can not be excluded. So, again brain morphology supports the clades Eunicoidea on the one-hand side and Oenonoidea on the other. The general brain anatomy of Dorvilleidae seems to be identical to Eunicidae and Onuphidae.

The general brain morphology seems not to be affected by the number of prostomial sensory organs because although differences in the shape of the brain might occur among the different families, they do not seem to be related to the number of prostomial sensory organs.

4.3. Phylogenetic Significance of Nerve Roots

Equal numbers and positions of nerve roots among the different families suggest a close relationship between Eunicidae and Onuphidae. The number of nerve roots is the same for the median antenna (4 main nerve roots) [13] as well as the palps (3 main nerve roots) in these taxa. In our study, the number of nerve roots of the lateral antennae differs among the families: while four nerve roots are present in Eunicidae, only two nerve roots were observed in *Hyalinoecia tubicola*. If this number is applicable to the complete clade of Onuphidae has to be evaluated by studying more specimens. It could be hypothesized that age effects may play a role, so that definitive innervation patterns may become apparent in the adult stage only. For example, Zanol [29] reported that juveniles of Eunicidae lack adult characters such as palps and lateral antennae and it could be suggested, that their innervation is not fully developed yet. Still, the specimen of *H. tubicola* we investigated seems to have already reached the adult stage as the remaining species studied. As the examined representatives of Onuphidae, Dorvilleidae lack minor nerves in the lateral antennae, so here also only two main nerve roots are present [29].

The palps of the examined Dorvilleidae differ in inner morphology as well as innervation from palps of other taxa as well as antennae of the same taxon. Dorvilleid palps are innervated by two distinct main nerves and possess a coelomic cavity and musculature extending from the palpophore to nearly the tip of the palpostyle. Whether this construction has a special function can only be speculated. Together with the exceptional innervation pattern, the differing inner palp morphology may support the phylogenetic placement of Dorvilleidae as a clade outside of Eunicoidea and Oenonoidea as proposed by Tilic et al. [2]. Whether Dorvilleidae forms a clade together with Histriobdellidae has still to be evaluated. According to Orrhage and Müller [13] and Purschke et al. [37], up to twelve different palp nerve roots can be present in Annelida, but until now no taxon studied possessed all twelve palp nerve roots [13,37]. The palp nerve roots we observed are the two thick main palp nerve roots originating from the drcc and the vrcc and some minor nerve roots (see Purschke et al. [37]).

4.4. Ecological Significance of Sensory Organs in Eunicidae

Antennae and palps are both presumably used to recognize prey in Eunicida. *Eunice aphroditois* (Pallas, 1788), commonly referred to as bobbit worm, for example, is an ambush predator and hides in the sediment—only its frontal part reaching out into the water column [46]. If potential prey gets accidentally in contact with the head appendages, the worm will capture it with its jaws. In endobenthic forms such as Oenonoidea head appendages and eyes are reduced in most species, which concurs with their lifestyle: in an endobenthic environment, the sensory function of head appendages, as well as optical senses, are of no great use. Generally, reductions are common in taxa-shifting lifestyles from epi- to endobenthic [47]. Small, interstitial eunicids as commonly found in Dorvilleidae (e.g., *Neotenotrocha*) may have reduced head appendages as well as innervation patterns due to progenesis [6,48]. Extreme differences in body sizes can frequently be observed in Eunicida, which are accompanied by different feeding modes and lifestyles: Small interstitial dorvilleids (1 mm) for example are microphagous "grazing" on algae and biofilms with their jaws (comparable to the radula of gastropods), while larger representatives, such as *Eunice aphroditois* (3 m), are macrophagous and predatory [49].

In the predatory Oenonoidea, the nuchal organ has a crucial function in detecting prey in this taxon. Von Haffner [11] hypothesised that the nuchal organ balances the missing sensory input of head appendages and eyes. In his studies, he theoretically separated the brain of Eunicida into different parts: a fore-brain, the mid-brain and a hind-brain. Based on this concept, he inferred, that the hind-brain in *Lysidice ninetta* and *Nematonereis unicornis* is bigger than in species bearing more head appendages. For *Scoletoma (Lumbrineris) fragilis*, he reported more prominent nuchal ganglia. We did not observe an extended posterior brain part or prominent nuchal commissures in any specimen.

While the surface of the nuchal cavity seems to be enlarged through foldings (x-shape) in the studied specimens of Oenonidae and Lumbrineridae, the receptive, ciliated surface is not enlarged when compared to the arc-shaped nuchal organs of Eunicoidea. Considering that in the examined specimens of Eunicoidea and Oenonoidea overall two ciliary patches of similar size are present, it could be assumed, that sensitivity of the nuchal organs is similar among these species. In contrast to this, the nuchal organs of *Dorvillea* spec. contain overall four ciliary patches—it shows a duplication of the common two ciliary patches. Here, the ventrally, laterally and dorsally located openings of the nuchal cavities may direct the water current directly onto the respective ciliary patch. This might enable *Dorvillea* spec. to detect the direction of, e.g., chemical cues ([39] for *Saccocirrus* species).

Lateral organs are present in every specimen of Eunicida and are regarded as characteristics belonging to the eunicidan ground pattern [4,37,50,51]. Generally, lateral organs are described as ciliated pits or papillae located between the noto- and the neuropodium in polychaetous annelids [4,52]. In our study, all lateral organs are based on the ventral side of the dorsal cirrus of the notopodium or an equivalent position in taxa lacking a notopodium, e.g., Lumbrineridae and Histriobdellidae. Additional differences in morphology of lateral organs occur in *Hyalinoecia tubicola* (Onuphidae) and *Arabella iricolor* (Oenonidae): here, the lateral organs are not only covered by cilia but also by microvilli. Moreover, the cilia of the lateral organ of *Ophryotrocha siberti* are longer than the ones in the remaining specimens studied. These characters support the hypothesis of species-specific morphologies of lateral organs in Eunicida. Hayashi and Yamane [53] reported differences in the histology of lateral organs of *Marphysa sanguinea* (Montagu, 1813) (Eunicidae) and *Scoletoma (Lumbrineris) longifolia* (Imajima and Higuchi, 1975) (Lumbrineridae) suggesting these to be species-specific.

4.5. High Metabolic Rate of the Brain

Regardless of the different lifestyles (endobenthic, epibenthic), the brain of the studied eunicidans exhibits the same indicators for a high metabolic rate. Blood vessels traverse the brain frequently indicating a high metabolic activity as typical for errant annelids. Moreover, three different kinds of neuronal somata surround the brain of the different species studied. Type one neurons are most abundant as typical for annelids [54]. Somata of type two and type three neurons generally contain densely arranged endoplasmatic reticula indicating a high metabolic rate [54]. Further, the high abundance of dense and lucent core vesicles in the brain neuropil indicates a high synaptic activity [34]. A high metabolic rate is a prerequisite for processing diverse sensory stimuli and for a predatory lifestyle. In all species examined these indicators (blood vessels, three types of neuronal somata, endoplasmatic reticula, dense and lucent core vesicles) can be found in the same abundancy and arrangement regardless of the number of prostomial sensory organs or lifestyle, which hint at a conserved status of these parameters.

4.6. Answers to Initial Questions

1. Is the brain morphology affected by different numbers and shapes of prostomial sensory organs? The gross brain morphology is not affected by the number and shape of sensory organs, but the microanatomy may be affected in terms of commissures and tracts (e.g., association tracts, optical commissures).
2. Do similar numbers and shapes of prostomial sensory organs hint at close phylogenetic relationships among different eunicidan taxa? Yes, if characters are mapped to the latest phylogenetic tree (see Table 2 and Figure 13, e.g., the clades Eunicoidea, as well as Oenonoidea, can be supported.
3. How can head appendages (antennae and palps) as well as buccal lips be differentiated? Differentiation is performed via innervation patterns: Antennae are solely innervated by nerves emanating from the drcc [13,14]. Palps are innervated by two or more nerve roots originating from the dorsal root of the circumoesophageal connective (drcc) and the ventral root of the circumoesophageal connective (vrcc) and their commissures [13]. Buccal lips are innervated solely by the ventral brain neuropil.

5. Conclusions

The gross morphology of the central nervous system is not influenced by the number of prostomial sensory organs and seems to be conserved among Eunicida when comparing only major components. When taking a closer look at more detailed morphologies, plasticity appears consolidating in apparent similarities between the studied specimens of Eunicidae + Onuphidae on the one-hand side and Oenonidae + Lumbrineridae on the other-hand side. Prostomial sensory organs shape the microanatomy of the brain in the studied representatives of Eunicidae and Onuphidae in a similar manner in terms of commissures and tracts. Further, the prostomial sensory organs such as antennae, palps, buccal lips, eyes and especially nuchal organs of both taxa occur in characteristic numbers and morphologies. In contrast to this, the studied representatives of Oenonidae and Lumbrineridae lack most of these prostomial sensory organs and the related commissures. Still, nuchal organs are present and based on a similar morphology, they could hint at a close relationship between Oenonidae and Lumbrineridae. Further, the presence of lateral neuropil areas and the dorso-ventrally extended brain shape could be used as further indicators for this relationship. When extrapolating the data gained in this study from species level to family rank, our data concur with the phylogeny of Eunicida proposing a sister group relationship of Eunicidae + Onuphidae (Eunicoidea) and Oenonidae + Lumbrineridae (Oenonoidea) [2]. In this phylogeny, Dorvilleidae builds a clade outside of Eunicoidae and Oenonoidea, which can be supported by the results gained in this study. If Dorvilleidae forms a clade together with Histriobdellidae has still to be evaluated. The dorsal nerves of *Arabella iricolor* could be homologized with nerves innervating antennae in Eunicidae and

Onuphidae and—supported by equivalent observations in Dorvilleidae by Zanol [29]—it could be hypothesized that a reduction of antennae might have taken place.

Supplementary Materials: The following supporting information can be downloaded at: https://www.mdpi.com/article/10.3390/jmse10111707/s1, Table S1. Specimens used in this study, sampling locality and methods applied. Overall 13 specimens of Eunicidae, 3 specimens of Onuphidae, 5 specimens of Dorvilleidae, 2 specimens of Oenonidae, 4 specimens of Lumbrineridae and 2 specimens of Histriobdellidae were used. *cLSM*: confocal laser scanning microscopy, *LM*: light microscopy, *TEM*: transmission electron microscopy, *SEM*: scanning electron microscopy; Table S2. Terminology of head appendages and buccal lips over time; Table S3. Links for downloading complete image stacks of the examined specimens.

Author Contributions: P.B. conceived the study. S.K. acquired and analyzed all data and wrote the manuscript. P.B. and T.B. reviewed and edited the draft. All authors have read and agreed to the published version of the manuscript.

Funding: The µCT-scanner was funded by the State of North Rhine-Westphalia and the Deutsche Forschungsgemeinschaft (INST 217/849-1 FUGG). The FEI Verios 460 L scanning electron microscope was funded by the State of North Rhine-Westphalia and the Deutsche Forschungsgemeinschaft (INST 217/784-1 FUGG).

Institutional Review Board Statement: Not applicable.

Informed Consent Statement: Not applicable.

Data Availability Statement: To facilitate future comparative analyses and to maintain data transparency, all image series of paraffin histology (AZAN + silver) are deposited in https://zenodo.org, accessed on 5 October 2022 (see Supplementary Material, Table S3). Complete image stacks can be downloaded using Zenodo_get (Völgyes and Lupton 2020). Physical sections are deposited in the histology collection of the Institute of Evolutionary Biology and Ecology of the University of Bonn as vouchers.

Acknowledgments: The authors thank the Station Marine de Concarneau for their support in collecting some of the examined animals. The authors thank Ekin Tilic for providing specimens of *Arabella iricolor* and *Ophryotrocha siberti*, and Conrad Helm for gently providing a specimen of *Hyalinoecia tubicola*. Further, we would like to thank Alexander Ziegler for his support with the µCT-scanner. We are very grateful to Gregor Kirfel from the Institute for Cell Biology in Bonn for his support with the FEI Verios 460 L scanning electron microscope.

Conflicts of Interest: The authors declare no conflict of interest.

References

1. Rouse, G.W.; Pleijel, F.; Tilic, E. *Annelida*; Oxford University Press: Oxford, UK, 2022; pp. 103–125.
2. Tilic, E.; Stiller, J.; Campos, E.; Pleijel, F.; Rouse, G.W. Phylogenomics resolves ambiguous relationships within Aciculata (Errantia, Annelida). *Mol. Phylogenetics Evol.* **2022**, *166*, 107339. [CrossRef]
3. von Palubitzki, T.; Purschke, G. Ultrastructure of pigmented eyes in Onuphidae and Eunicidae (Annelida: Errantia: Eunicida) and its importance in understanding the evolution of eyes in Annelida. *Zoomorphology* **2020**, *139*, 1–19. [CrossRef]
4. Rouse, G.; Pleijel, F. *Polychaetes*; Oxford University Press: Oxford, UK, 2001. [CrossRef]
5. Ehlers, E. *Die Borstenwürmer Nach Systematischen und Anatomischen Untersuchungen Dargestellt*; Wilhelm Engelmann: Leipzig, Germany, 1868.
6. Struck, T.H.; Purschke, G.; Halanych, K.M. Phylogeny of Eunicida (Annelida) and exploring data congruence using a partition addition bootstrap alteration (PABA) approach. *Syst. Biol.* **2006**, *55*, 1–20. [CrossRef]
7. Zanol, J.; Carrera-Parra, L.F.; Steiner, T.M.; Amaral, A.C.Z.; Wiklund, H.; Ravara, A.; Budaeva, N. The Current State of Eunicida (Annelida) Systematics and Biodiversity. *Diversity* **2021**, *13*, 74. [CrossRef]
8. Paxton, H. Phylogeny of Eunicida (Annelida) based on morphology of jaws. *Zoosymposia* **2009**, *2*, 241–264. [CrossRef]
9. Kielan-Jaworowska, Z. Scolecodonts versus jaw apparatuses. *Lethaia* **1968**, *1*, 39–49. [CrossRef]
10. Hartmann-Schröder, G. *Polychaeta*, 2nd ed.; Fischer: Jena, Germany, 1996.
11. von Haffner, K. Über die Abhängigkeit des Gehirnbaues von den Sinnesorganen: Vergleichende Untersuchungen an Euniciden (Polychaeta). *Zool. Jb. Anat* **1962**, *80*, 159–212.
12. Hanström, B. Das zentrale und periphere Nervensystem des Kopflappens einiger Polychäten. *Z. Morphol. Ökologie Tiere* **1927**, *7*, 543–596. [CrossRef]

13. Orrhage, L.; Müller, M.C.M. Morphology of the nervous system of Polychaeta (Annelida). *Hydrobiologia* **2005**, *535–536*, 79–111. [CrossRef]
14. Orrhage, L. On the Innervation and Homologues of the Anterior End Appendages of the Eunicea (Polychaeta), with a Tentative Outline of the Fundamental Constitution of the Cephalic Nervous System of the Polychaetes. *Acta Zool.* **1995**, *76*, 229–248. [CrossRef]
15. Hessling, R.; Purschke, G. Immunohistochemical (cLSM) and ultrastructural analysis of the central nervous system and sense organs in *Aeolosoma hemprichi* (Annelida, Aeolosomatidae). *Zoomorphology* **2000**, *120*, 65–78. [CrossRef]
16. Beckers, P.; Helm, C.; Bartolomaeus, T. The anatomy and development of the nervous system in Magelonidae (Annelida)—Insights into the evolution of the annelid brain. *BMC Evol. Biol.* **2019**, *19*, 173. [CrossRef]
17. Monro, C.C.A. Polychaete worms. In *Discovery Reports*; University Press: Cambridge, UK, 1931; Volume II, pp. 3–218.
18. Aiyar, R.G. On the Anatomy of *Marphysa Gravelyi* Southern. *Rec. Zool. Surv. India* **1933**, *35*, 287–323. [CrossRef]
19. Fauchald, K. *The Polychaete Worms: Definitions and Keys to the Orders, Families and Genera*; Natural History Museum of Los Angeles County, Science Series: Los Angeles, CA, USA, 1977.
20. Fauchald, K.; Rouse, G.W. Polychaete systematics: Past and present. *Zool. Scr.* **1997**, *26*, 71–138. [CrossRef]
21. Heider, K. Über Eunice. Systematisches, Kiefersack, Nervensystem. *Z. Für Wiss. Zool.* **1925**, *125*, 55–90.
22. von Haffner, K. Über den Bau und den Zusammenhang der Wichtigsten Organe des Kopfendes von Hyalinoecia Tubicola Malmgren (Polychaeta, Eunicidae, Onuphiidae) mit Berücksichtigung der Gattung Eunice. *Zool. Jahrbücher Abt. Anat. Ontog. Tiere* **1959**, *77*, 113–192.
23. von Haffner, K. Untersuchungen und Gedanken über das Gehirn und Kopfende von Polychaeten und ein Vergleich mit den Arthropoden. *Mitt. Zool. Staatsinst. Zool. Mus. Hambg.* **1959**, *37*, 1–29.
24. Paxton, H. Generic revision and relationships of the family Onuphidae (Annelida: Polychaeta). *Rec. Aust. Mus.* **1986**, *38*, 1–74. [CrossRef]
25. Orensanz, J.M. The eunicemorph polychaete annelids from Antarctic and Subantarctic seas: With addenda to the eunicemorpha of Argentina, Chile, New Zealand, Australia, and the southern Indian Ocean. *Biol. Antarct. Seas XXI* **1990**, *52*, 1–183. [CrossRef]
26. Binard, A.; Jeener, R. Morphologie du lobe préoral des polychètes. *Recl. De L'Inst. Zool. Torley-Rousseau* **1928**, *2*, 177–240.
27. Gustafson, G. Anatomische Studien über die Polychaeten-Familien Amphinomidae und Euphrosynidae. Ph.D. Thesis, Almqvist & Wiksell, Stockholm, Sweden, 1930.
28. Åkesson, B. The Embryology of the Polychaete *Eunice kobiensis*. *Acta Zool.* **1967**, *48*, 142–192. [CrossRef]
29. Zanol, J. Homology of prostomial and pharyngeal structures in eunicida (Annelida) based on innervation and morphological similarities. *J. Morphol.* **2010**, *271*, 1023–1043. [CrossRef] [PubMed]
30. Pruvot, G.; Racovitza, E.-G. Matériaux pour la faune des Annélides de Banyuls. *Arch. Zool. Expérimentale Générale* **1895**, *3*, 339–492.
31. Fauvel, P. Faune de France: 5 Polychètes Errantes. Fédération Française des Sociétés de Sciences Naturelles: Paris, France, 1923; pp. 4–491.
32. Müller, M.C.M. Polychaete nervous systems: Ground pattern and variations—cLS microscopy and the importance of novel characteristics in phylogenetic analysis. *Integr. Comp. Biol.* **2006**, *46*, 125–133. [CrossRef] [PubMed]
33. Beckers, P.; Helm, C.; Purschke, G.; Worsaae, K.; Hutchings, P.; Bartolomaeus, T. The central nervous system of Oweniidae (Annelida) and its implications for the structure of the ancestral annelid brain. *Front. Zool.* **2019**, *16*, 6. [CrossRef] [PubMed]
34. Beckers, P.; Pein, C.; Bartolomaeus, T. Fine structure of mushroom bodies and the brain in *Sthenelais boa* (Phyllodocida, Annelida). *Zoomorphology* **2021**, *3*, 579. [CrossRef]
35. Beckers, P.; Müller, C.; Wallnisch, C.; Bartolomaeus, T. Getting two birds with one stone: Combining immunohistochemistry and Azan staining in animal morphology. *J. Biol. Methods* **2021**, *8*, e153. [CrossRef]
36. Richter, S.; Loesel, R.; Purschke, G.; Schmidt-Rhaesa, A.; Scholtz, G.; Stach, T.; Vogt, L.; Wanninger, A.; Brenneis, G.; Döring, C.; et al. Invertebrate neurophylogeny: Suggested terms and definitions for a neuroanatomical glossary. *Front. Zool.* **2010**, *7*, 1–49. [CrossRef]
37. Purschke, G.; Bleidorn, C.; Struck, T. Systematics, evolution and phylogeny of Annelida—A morphological perspective. *Mem. Mus. Vic.* **2014**, *71*, 247–269. [CrossRef]
38. Malaquin, A. Recherches Sur Les Syllidiens: Morphologie, Anatomie, Reproduction, Développement. Danel, L., Ed.; Lille, France, 1893; pp. 3–477. Available online: https://www.biodiversitylibrary.org/bibliography/99327 (accessed on 5 October 2022).
39. Purschke, G. Ultrastructure of Nuchal Organs in Polychaetes (Annelida)—New Results and Review. *Acta Zool.* **1997**, *78*, 123–143. [CrossRef]
40. Ax, P. *Multicellular Animals: The Phylogenetic System of the Metazoa*; Springer Science & Business Media: Berlin/Heidelberg, Germany, 2000.
41. Strausfeld, N.J.; Hansen, L.; Li, Y.; Gomez, R.S.; Ito, K. Evolution, Discovery, and Interpretations of Arthropod Mushroom Bodies. *Learn. Mem.* **1998**, *5*, 11–37. [CrossRef]
42. Loesel, R.; Heuer, C.M. The mushroom bodies—Prominent brain centres of arthropods and annelids with enigmatic evolutionary origin. *Acta Zool.* **2010**, *91*, 29–34. [CrossRef]
43. Heuer, C.M.; Müller, C.H.; Todt, C.; Loesel, R. Comparative neuroanatomy suggests repeated reduction of neuroarchitectural complexity in Annelida. *Front. Zool.* **2010**, *7*, 13. [CrossRef] [PubMed]
44. Bullock & Horridge. *Structure and Function of the Nervous System of Invertebrates*; Freeman: San Francisco, CA, USA, 1965.

45. Racovitza, E.G. Le lobe céphalique et l'encéphale des Annélides Polychètes (Anatomie, Morphologie, Histologie). *Arch. Zool. Expérimentale Générale* **1896**, *3*, 133–343.

46. Huisman, J.; Dixon, R. Northern exposure: Marine life of the Kimberley. *Landscope* **2011**, *26*, 10–16.

47. Purschke, G.; Nowak, K.H. Ultrastructure of pigmented eyes in Dorvilleidae (Annelida, Errantia, Eunicida) and their importance for understanding the evolution of eyes in polychaetes. *Acta Zool.* **2015**, *96*, 67–81. [CrossRef]

48. Struck, T.H.; Golombek, A.; Weigert, A.; Franke, A.; Westheide, W.; Purschke, G.; Bleidorn, C.; Halanych, K.M. The evolution of annelids reveals two adaptive routes to the interstitial realm. *Curr. Biol.* **2015**, *25*, 1993–1999. [CrossRef]

49. Purschke, G.; Tzetlin, A.B. Dorsolateral ciliary folds in the polychaete foregut: Structure, prevalence and phylogenetic significance. *Acta Zool.* **1996**, *77*, 33–49. [CrossRef]

50. Beesley, P.L.; Ross, G.J.B.; Glasby, C.J.; Study, A.B.R. *Polychaetes & Allies: The Southern Synthesis*; CSIRO Publishing: Clayton, Australia, 2000.

51. Budaeva, N.; Schepetov, D.; Zanol, J.; Neretina, T.; Willassen, E. When molecules support morphology: Phylogenetic reconstruction of the family Onuphidae (Eunicida, Annelida) based on 16S rDNA and 18S rDNA. *Mol. Phylogenetics Evol.* **2016**, *94*, 791–801. [CrossRef] [PubMed]

52. Purschke, G. Sense organs in polychaetes (Annelida). *Hydrobiologia* **2005**, *535–536*, 53–78. [CrossRef]

53. Hayashi, I.; Yamane, S. Further Observation of a Recently Found Sense Organ in Some Euniciforms, with Special Reference to *Lumbrineris Longifolia* (Polychaeta, Lumbrineridae). *Bull. Mar. Sci.* **1997**, *60*, 564–574.

54. Beckers, P.; Tilic, E. Fine structure of the brain in Amphinomida (Annelida). *Acta Zool.* **2021**, *11*, 2860. [CrossRef]

MDPI
St. Alban-Anlage 66
4052 Basel
Switzerland
Tel. +41 61 683 77 34
Fax +41 61 302 89 18
www.mdpi.com

Journal of Marine Science and Engineering Editorial Office
E-mail: jmse@mdpi.com
www.mdpi.com/journal/jmse